WHAT IS TUBERCULOSIS ?

WHAT IS TUBERCULOSIS?

ITS CONTROL AND PREVENTION

By

MILOSH SEKULICH, M.D. (BELGRADE)

Former Chief Physician of the Municipal Hospital of Belgrade

With a Preface by

H.M. QUEEN MARIA OF YUGOSLAVIA

Edited by

MARK CLEMENT

LONDON:

RESEARCH BOOKS LIMITED

in association with

WILLIAM HEINEMANN (MEDICAL BOOKS) LIMITED

1944

Printed in Great Britain by Love & Malcomson, Limited, Redhill

CONTENTS

PART I

WHAT IS TUBERCULOSIS?

Preface by H.M. Queen Maria of Yugoslavia

Introduction, *p.* 1

Definition of Tuberculosis, *p.* 5

The cause of Tuberculosis, *p.* 5

Characteristics and varieties of Bacillus Tuberculosis, *p.* 6

Tuberculosis and Heredity, *p.* 7

Tuberculin Reactions, *p.* 7

Significance of Tuberculin Reactions, *p.* 8

Tuberculosis Infection, *p.* 9

Sources of Infection, *p.* 10

Typical cases of Infection, *p.* 12

Changes taking place in an organism infected with Bacillus Tuberculosis, *p.* 14

First symptoms of Tuberculosis of the Lungs, *p.* 15

Diagnosis, *p.* 19

Mass Radiography and Tomography, *p.* 21

Should a consumptive be told the truth? *p.* 23

Tuberculosis is a social disease, *p.* 24

Social causes of Tuberculosis, *p.* 25

The economic factor as a cause of Tuberculosis, *p.* 27

Tuberculosis and the home, *p.* 28

Food and Tuberculosis, *p.* 29

Tuberculosis and alcoholism, *p.* 30

Tuberculosis and mode of life, *p.* 30

Social Pathology of Tuberculosis, *p.* 31

The spread of Tuberculosis, *p.* 34

The fight against Tuberculosis in rural districts, *p.* 38

Mortality from Tuberculosis in Yugoslavia, *p.* 41

Mortality from Tuberculosis in England and Wales, *p.* 42

PART II

PREVENTION AND TREATMENT

How to prevent infection, *p.* 45

Preventing Tubercle Bacilli entering the body, *p.* 46

Milk and Pasteurisation, *p.* 47

Disinfection of sputum, *p.* 49

Disinfection of table ware, *p.* 50

Disinfection of clothing and linen, *p.* 50

Disinfection of books and letters, *p.* 51
Disinfection of rooms, *p.* 51
Disinfection of floors, *p.* 51
Disinfection of furniture, *p.* 52
Calmette - Guerin Vaccine (BCG), *p.* 52
Protection of children, *p.* 54
Tuberculosis and marriage, *p.* 56
Building up resistance to Tuberculosis, *p.* 57
Education and propaganda, *p.* 37
Importance of scientific work, *p.* 58
Individual Treatment of Tuberculosis of the Lungs, *p.* 58
Aims and Methods of Individual Treatment, *p.* 58
Hygienic add Dietetic Régime, *p.* 60
Fresh Air, *p.* 61
Climatic Treatment, *p.* 62
Rest, *p.* 63
Exercise and Recreation, *p.* 64
Sports and Work, *p.* 66
Hydrotherapy, *p.* 66
Heliotherapy, *p.* 67
Diet, *p.* 67
Special Methods in the treatment of Pulmonary Tuberculosis, *p.* 71
Immunotherapy, *p.* 71
Chemotherapy, *p.* 72
Collapse Therapy, *p.* 72
Artificial Pneumothorax (A.P.), *p.* 73
Oleothorax, *p.* 74
Pneumolysis, *p.* 74
Phrenic Nerve Operations, *p.* 75
Phrenic Alcoholisation, *p.* 76
Thoracoplasty, *p.* 76
Extrapleural Pneumolysis, *p.* 77
Closed suction drainage, *p.* 78
Combination of methods, *p.* 78
Treatment of Tuberculous Diabetics, *p.* 78
Medicinal Treatment, *p.* 79
Treatment of Surgical Tuberculosis, *p.* 80
Institutions for Treatment and Prevention, *p.* 81
The Night Rest Centre, *p.* 88
The Day Rest Centre, *p.* 88
The Hospital, *p.* 89
The Sanatorium, *p.* 89
The Isolation Home, *p.* 90
The Convalescent Home, *p.* 90
The Colony, *p.* 90
Papworth Village Settlement, *p.* 91
Is Pulmonary Tuberculosis curable ? *p.* 94
What we understand by cure and how to prove it, *p.* 95
Is Tuberculosis as a social disease curable ? *p.* 95

PREFACE

Tuberculosis even in peace-time is one of the diseases that has to be fought with all the energy and means at our disposal. In war-time it is more prevalent but it only shows itself in its most virulent form in the years that follow after war. You will find that even here in Great Britain there is more tuberculosis than there was in peace-time. In my own country we have had a very hard battle against this scourge and we know that on our return to Yugoslavia we shall have to start all over again to do all that is in our power to conquer this enemy.

From reports that I have received from Yugoslavia I understand that there are about 400,000 children who are unprotected, many of whom are orphans and of these 60% are already infected. This proves that everything that is humanly possible must be done to protect all peoples from these heartrending conditions.

Tuberculosis *can* be cured and prevented by doctors and scientists but *only* with the good-will and co-operation of the entire population. Every man, woman, and child must make it their duty to help in this fight to free the world of one of its most deadly perils.

Maria

PUBLISHER'S NOTE

Thanks are due to the Yugoslav Association of St. Andrija which has made the publication of this book possible.

WHAT IS TUBERCULOSIS?

INTRODUCTION

After this war there will be chaos all over Europe. The destruction of a great number of towns and villages will have made millions of people homeless. The shortage of workers caused by mass movements of populations, the calling up of men to work for the Axis war machine in Germany, Italy and the satellite states, as well as in their own occupied countries, the segregation of many prisoners of war, the imprisonment of people in concentration camps, and the taking of hostages, all this will cause starvation and extreme poverty. If things are going from bad to worse in Axis countries as the war progresses what must be the situation in occupied countries robbed of foodstuffs and other necessities of life?

According to various reliable reports which have reached Allied Governments, the situation in occupied Europe is already desperate and is getting worse and worse. In all these reports great stress is laid on the fact that malnutrition is prevalent everywhere and as a result tuberculosis is spreading like a plague. When the fighting ends the situation will be inconceivably difficult.

As a result of these circumstances the death-rate among the civilian population in Europe has risen enormously during the war. This terrible state of affairs is not merely temporary but will last for many years after the war. Epidemics and the incidence of various diseases will not end with the fighting but will continue to increase.

Statistics show that for four years after the last war more people died from starvation and disease, which could have been prevented, than were killed in action. It is clear that the people of Europe will need, first of all, food, clothes and medical supplies. Some of the essential necessities during the first period of relief could be taken from the existing armies. But besides this, we must start accumulating stores of equipment and foodstuffs now. Our experience after the last war fully justifies the demand for timely and adequate measures.

We recall the great work done then by the British and American Relief Organisations in Serbia, and the good use the starving and ill-clad people made of what they were given by the kind British and American relief workers.

During the first period we may need only those medical supplies, food and clothes which will bring immediate relief—but during the second period it will be imperative to build up and reorganise all the medical services and provide them with all the necessary supplies for the purpose of social rehabilitation.

The political, economic, social and cultural differences which divided the peoples of Europe before and during the war will persist for some time after the war. In such a situation a well-run relief organisation will provide appropriate remedies for the many ills afflicting occupied Europe. The medical relief reconstruction organisation must be non-political. Every unit must be run by doctors whose one aim is to heal the sick and serve humanity, free from any control by the military authorities, political parties or factions.

The problem of general reconstruction is a matter concerning many experts. In the present work I will confine myself to my own subject which is *medical relief reconstruction based on the fight against tuberculosis.* If we can win this fight we shall have overcome the major part of the problem of medical reconstruction.

Various diseases attack different districts at different seasons, for example, malaria in the summer. *Tuberculosis is a most dangerous and widespread disease; its incidence has increased during the war, and will continue to increase still more for many years afterwards mainly owing to privations and under-nourishment.*

Even in Great Britain where conditions are better than anywhere in Europe tuberculosis is increasing. In other parts of Europe the situation is much more complex. In the neutral states, the increase of tuberculosis is about the same as in Great Britain while in Germany and her satellite states the increase is greater. But in occupied countries the situation is very grave indeed. In the first group the causative factors are on the increase; in the second group they are increasing more rapidly, and in the third group they are rampant.

Everywhere this situation will be aggravated towards the end of the war and will become even worse after the cessation of hostilities.

It is of paramount importance that in Great Britain to-day the responsible authorities should take immediate steps to arrest the progress of tuberculosis and make plans to combat the new causes which are developing, and which will multiply during the last period of the war. It is important that in conjunction with the Allies we should plan for the occupied countries not only for an emergency period after the war but for many years ahead.

The British Ministry of Health has set a fine example by introducing mass radiography in its new campaign against tuberculosis. Thus the British authorities are working on medical reconstruction even before the end of the war, and they have singled out tuberculosis as one of their main objectives. This supports our view that the fight against tuberculosis is the most important branch of medical reconstruction to be undertaken in Europe. A campaign against tuberculosis is the first essential step in a policy for public health. The aim of such a campaign is not only to detect tuberculosis in those who do not know they are suffering from it, or to detect it in its early stages and cure the victims, but rather to build up resistance against disease in general. In fighting tuberculosis we are doing the most important part of the work of general medical reconstruction.

The problem of tuberculosis is of such magnitude that it cannot be solved by doctors alone who will need the help of the State as well as that of hosts of social workers. Tuberculosis is not only a disease which everyone fears, it is one of the deadliest enemies of mankind, and everyone should be enlightened about its destructive power. It is also essential that we should know the best methods to fight against it.

Modern medicine has all the necessary resources to wage a successful fight against Tuberculosis, and scientific research has taught us that this disease, while attacking the individual, is primarily a social disease which has its roots in the bad conditions under which people live in modern times.

Before the war congested towns created a demand for more hospitals and it should be realised that after the war there will be not only congested towns, but also many localities and even whole regions entirely destroyed.

It is important to make people understand the problem and nature of tuberculosis so that they may play their part in keeping fit, and if the disease should appear its manifestations will be recognised and help will be sought without delay. Skilled help at the first sign of trouble achieves the best results, and if the disease is diagnosed in its early stages it can be cured more easily and rapidly.

It is equally important to stress that Tuberculosis in its various aspects should be understood not only by those suffering from it, and by people coming into contact with them, but also by the general public, and in particular by those who will be helping to fight it, as well as those planning medical relief reconstruction.

In order that the man in the street may understand the nature of the problem and play his part in helping to solve it, the present work has been written which deals briefly with every aspect of the prevention and treatment of tuberculosis.

As the reader will see the organisation to fight tuberculosis can be easily adapted for general medical relief reconstruction.

DEFINITION

Pulmonary tuberculosis was known to the Greeks and was considered contagious by Hippocrates, the Father of Medicine.

In the seventeenth century it was noticed that people who died from consumption had small grey knots or nodules in their lungs. The Latin word for knot is tuberculum, hence the name tuberculosis.

Tuberculosis was generally called consumption, because the organism wasted away and the Greeks expressed this idea with full reason by the word "Phthisis."

For a long time the cause of tuberculosis was unknown in spite of the fact that doctors were familiar with the nature and course of the disease.

THE CAUSE OF TUBERCULOSIS

In 1865 a Frenchman, Dr. Villemin, announced to the Paris Academy of Medicine that tuberculosis was a contagious disease. To prove his theory he transplanted a piece of diseased tissue from the lung of a man who had died from tuberculosis into the lung of a healthy animal. After a time the animal became ill and died. This also happened when he transplanted a piece of the lung of a tuberculous animal into the lung of a healthy one. He then noticed that the animal's lungs had become affected in the same way as the consumptive's lungs. Villemin made many experiments with the same results. His ideas were considered fantastic at the time and caused much discussion. He persisted in his views in spite of his failure to discover the germ. It was at last discovered in 1882 by a German doctor, Robert Koch, who demonstrated that the germ is always found in the diseased tissue, or in pus, or mucus from it. He then found the germ in the sputum of a consumptive. He demonstrated his discovery by staining the germ so that it could be seen through a microscope. He then succeeded in making a culture from the sputum of a consumptive with which he inoculated animals and produced the same disease in them. This is called experimental tuberculosis. It was thus made clear that tuberculosis was a contagious disease caused by a germ, called the tubercle bacillus. Koch's discovery was an outstanding achievement and made him a universal benefactor. From this moment tuberculosis research made rapid progress.

CHARACTERISTICS OF BACILLUS TUBERCULOSIS

The tubercle bacillus is shaped like a rod and measures about 3/1000 of a millimetre. It is made visible under the microscope by special staining methods. The tubercle bacillus is very resistant and can live for a long time outside the body especially in damp and dark places. In sunshine and well-lit places it loses its virulence but in snow and ice it can survive. In boiling water it is killed in a few minutes. It is protected by a fatty capsule which resists the action of acids.

VARIETIES OF BACILLUS TUBERCULOSIS

There are four principle types: 1, Human. 2, Bovine. 3, Avian (birds). 4, Piscine (fishes).

Human beings are generally infected by the human type while the bovine type is a common source of infection in children. It is estimated that 25% of the deaths from all forms of tuberculosis can be attributed to the bovine bacillus. Neither the avian nor the piscine type is found in man.

There has been much discussion about "virus filtrans," a very small germ which cannot be seen by any magnifying apparatus and which is said to cause tuberculosis. This germ is so small that it can pass through the finest filter, through which other germs including bacillus tuberculosis cannot pass. Some research workers state that they have produced experimental tuberculosis with this very small germ, and if it can be proved that it does actually exist then the present conceptions of tuberculosis, especially concerning heredity, may have to be modified.

TUBERCULOSIS AND HEREDITY

It is considered to-day that a healthy child is born of a consumptive mother. If the child is taken away from its mother at birth and placed in healthy surroundings it will be saved from tuberculosis. Only in exceptional cases can it be infected in its mother's womb. The most common source of infection is unhealthy surroundings. Hence the origin of the saying: "There is tuberculosis in the family." All modern research so far shows that tuberculosis is not hereditary. To illustrate this point one method of examination will be given, namely the reaction of the organism to tuberculin.

TUBERCULIN REACTIONS

The principle of biological tuberculin skin reaction is as follows: If we inoculate a healthy guinea pig with tubercle bacilli a swelling will appear and later a wound which will remain until the death of the animal. In the case of a tuberculous guinea pig the wound will be small and will soon disappear. Thus we see that the infected organism reacts in a different way. This different reaction goes under the name of "allergy." The organism reacts allergically if we use dead bacilli or a culture made of them called tuberculin. An inoculation of tuberculin into a healthy animal will not produce any reaction but a very small dose into an infected animal produces a strong reaction.

To ascertain if the organism has been infected or not we use the Pirquet and Mantoux tests named after their discoverers. For these tests we use tuberculin, a culture of dead tuberculosis bacilli containing a certain amount of medium in which they are grown. To obtain a tuberculin reaction (Pirquet) we use a special lancet with which the skin is scarified (scratched) in three different places, two or three inches apart. The middle scratch serves as the control; on the other two scratches one drop of tuberculin is placed. Twenty to thirty minutes are then allowed for the tuberculin to be absorbed. The reaction manifests itself within 24 hours. In some cases reaction is delayed for two or three days. The reaction is positive if the skin on the tuberculin scratch is red or sometimes swollen, while the control scratch remains unchanged.

The Mantoux test is carried out by means of a diluted solution of tuberculin. As a rule three solutions of different concentrations are prepared. They are injected into the skin of the forearm in three places, one-tenth of the above solutions, two inches apart. For the control one-tenth of physiological solution (salt solution) is used. The result is seen 24-48 hours later. Reaction is positive when the skin at the point of injection with tuberculin is red or sometimes swollen, while the skin at the point of control remains unchanged. The Pirquet test is generally performed first. If it is negative then the Mantoux test is carried out.

Another tuberculin test is known as the Moro test. For this test an ointment containing tuberculin is applied by inunction (1 c.mm.). As a rule a reaction appears within 24 hours. If it is positive an erythema is produced but its appearance may be delayed for two or three days.

The organism is not infected when the reactions are negative three times in succession over a period of fortnightly intervals. The majority of doctors are satisfied with one Pirquet test, and if it is negative, the Mantoux test is carried out.

SIGNIFICANCE OF TUBERCULIN REACTIONS

An organism which has never been infected with tuberculosis gives a negative reaction. If it has been infected the reaction is positive. A positive reaction means a certain degree of allergy (resistance). The tuberculin reaction again becomes negative when resistance is overcome by severe consumption.

A new-born baby always shows a negative reaction. This means that a child is born free from tuberculous infection. This is one of the proofs that tuberculosis is not hereditary. A positive reaction in an infant means the presence of tuberculosis.

TUBERCULOSIS INFECTION

MODES OF INFECTION; HOW BACILLUS TUBERCULOSIS ENTERS THE BODY

There are two ways by which the tubercle bacillus enters the body: by inhalation and by ingestion. In exceptional cases the bacilli enter through skin wounds, mucous membranes of the eyes and other organs. The most general way is by direct inhalation or by way of the mouth and throat, the bacilli entering through contaminated foods or by other means. This is the reason why the lungs are generally the first organs to be attacked and tuberculosis of the lung is the most common form of the disease.

To demonstrate this there are two conclusive experiments. If a guinea pig is put in a box and a duster containing tubercle bacilli is shaken over it, or if water containing bacilli is sprayed over the guinea pig, it will be infected and die. A post-mortem examination will show tuberculosis in the lungs. Infection through ingestion is as common as by inhalation but not so dangerous to adults as to infants. The bacilli pass from the digestive organs through the mucous membranes into the lymphatic vessels and into the lungs in both adults and infants.

The secretions of the digestive organs reduce the virulence of the bacilli, but these secretions are more potent in adults than in infants. This is why adults very rarely contract tuberculosis of the bovine type of bacillus. But infants whose gastric secretions are not so potent and whose main diet is milk contract tuberculosis as a result of drinking infected milk from a diseased cow. This explains why tuberculosis of the abdominal glands is often found in children

SOURCES OF INFECTION

The most dangerous source of infection is a patient with "open" tuberculosis of the lungs. If tubercule bacilli are found in the sputum of a consumptive he is said to suffer from "open" tuberculosis. In the sputum of a consumptive particles of destroyed lung tissue and an enormous number of bacilli are found. It is obvious that if a patient with "open" tuberculosis spits on the floor or in the street instead of into a pocket spittoon which can be closed, he spreads the disease.

It has been shown that when a person coughs particles of sputum are projected from 60–80 centimetres to one metre. If a glass plate is held in front of a consumptive's mouth and he is asked to cough, drops collect on the plate containing bacilli which can be seen through the microscope. Even if a consumptive spits into a pocket spittoon it is possible for some drops to fall on the floor or on articles of furniture, books, etc. These drops eventually dry up and are mixed with dust and so spread in the atmosphere.

Thus it becomes clear how the disease is spread by people spitting in public places ('buses, restaurants, parks, streets, etc.).

If a consumptive holds his hand in front of his mouth, if he coughs and then pats a child's head or shakes hands with another person he helps to spread the disease. It is most dangerous for a consumptive to kiss anyone especially a small child. Infection is also spread by dishes and table ware which have been used by a consumptive. There should be a rule never to allow anyone to use vessels, cups, or utensils which have been used by other persons. Flies are great carriers of germs as they settle on expectorations as well as on foodstuffs.

It is most dangerous to sleep in the same room or bed as a consumptive.

Tubercle bacilli can also be spread from tuberculous wounds or dressings or from the urine of a patient suffering from tuberculosis of the kidneys, or from the excretions of a person with tuberculosis of the intestines.

Tuberculosis can be spread over a great distance by means of a letter or other articles which have been coughed upon by a consumptive, or by a loaf of bread or any other foodstuff which has been handled by a consumptive.

Tuberculosis can be spread by a consumptive dairymaid milking the cows and handling the pail into which the milk is poured. It can also be spread by people drinking milk from a tuberculous cow. Hence the necessity of Pasteurisation.

As we have seen, the most dangerous source of infection is a person with "open" tuberculosis of the lungs who lives with other people. For example, a nursing mother with tuberculosis will pass the disease on to her child. Grandparents and other relatives who play with children can infect them. It is seldom introduced into a house by guests or by domestic animals.

The most likely places of infection are flats, factories and offices, schools, recreation grounds, public libraries, railway stations, restaurants and sometimes preventoria and convalescent homes and very rarely hospitals and sanatoria.

Objects by means of which tubercle bacilli can be spread are dolls and other toys, handkerchiefs, books, etc.

Sometimes people wonder why a whole family dies of tuberculosis with the exception of the grandparents who survive. The reason is that the grandparents are carriers. They play with the children, they kiss them, they cough, they taste food before giving it to the children and thus spread the disease. The most dangerous source of infection for children is milk. It is estimated that about 40% of the cattle in England and Wales are infected with tuberculosis.

This means that nearly all bulked milk coming into cities and towns is infected and therefore it should always be boiled before being consumed. It is estimated that about 2,000 human deaths in England and Wales are due to bovine tuberculosis conveyed by milk.

In a word, wherever tuberculosis is found it can be caught by anyone.

This point must be stressed and understood if we are to fight successfully against tuberculosis.

TYPICAL CASES OF INFECTION

The following examples of infection are taken from actual cases.

(1) A young mother was a tuberculous sufferer without being aware of it. She feared the disease so much that her doctor and husband told her that she had catarrh. She went to a sanatorium thinking she was being treated for catarrh. But after a time she became worried about her children and wanted to return home. The doctor knew that if he told her the truth it would be bad for her from a psychological point of view and she would get worse. He therefore allowed her to return home. The husband tried to keep the children away from her, but she insisted on seeing them for a little while every day saying that if she was not allowed to see them she must surely be suffering from tuberculosis. She became worse and died. By that time she had, of course, infected her children. The son died from tuberculous meningitis soon after and the daughter spent many years in a sanatorium.

(2) A mother decided to take her child who had a cold to the seaside as she thought that the change of air would do it good. She asked the child's nurse to come with them. The nurse refused saying that she had noticed before that the sea air did not agree with her. The mother suggested that she should consult the family doctor and if he said the nurse could go to the seaside without ill-effects then they

would go. The nurse was examined and it was found she had open tuberculosis. The doctor suggested that he should examine the child to ascertain if it had been infected. He did so and found that the cold was in fact tuberculosis. The child and mother went to a mountain resort and in time the child got well. The nurse foolishly refused treatment and took a job as a waitress. By this time she showed unmistakable signs of tuberculosis and could not help spreading the disease.

(3) There was a poor family of seven people living in very sordid surroundings. The father was gravely ill with tuberculosis. When the doctor visited their one-room home he found the living conditions appalling. In one corner of the room were a stove and a sink while the beds were set around the walls. The younger children were playing on the floor. The eldest, a girl of ten, was trying to stop her father's coughing with a towel. He was spitting blood all over the room and some of it was found on the children's toys. It was impossible to find a place for the father in an isolation home and eventually he died. Soon after the children became ill and died, one after the other.

(4) A man, his wife and two children were living in good circumstances. The father caught tuberculosis and the doctor suggested that the children should be sent away at once. The mother, however, decided to send her son to a boarding school and kept the daughter with her. In time the father died. It was noticed later that the daughter was always subject to colds. These colds persisted right up to the time of marriage and after she had a child it was discovered that these colds were in fact tuberculosis. She had treatment and in time she was cured.

(5) A young man was seriously ill with "galloping consumption." He had to be cared for at home as there was no room in the isolation home. His sister nursed him. Some time after he died while she became ill and developed a dry cough. She consulted the doctor and told him that

she sweated at night and felt weak. The doctor found tuberculosis. She was treated and was finally cured.

(6) Two school friends of healthy parents went to an elementary school in a small town. They took extra lessons at the home of their young woman teacher. The teacher had a cough and when one of the children became ill he was found to be suffering from tuberculosis. The doctor looked for the source of infection and discovered that the teacher was suffering from tuberculosis. She died later but the child got well.

(7) In a healthy village a school teacher was suffering from tuberculosis. She carried on her work for about two years and during that time a large number of children became ill with tuberculous meningitis. When the doctor looked for the source of infection he soon traced it to the teacher.

The above cases are typical and could happen anywhere. They illustrate the point that tuberculosis is a social disease.

CHANGES TAKING PLACE IN AN ORGANISM INFECTED WITH BACILLUS TUBERCULOSIS

When tubercle bacilli enter the human body they settle in an organ and attack it. The most commonly attacked organs are the lungs but any organ can be infected : digestive organs, kidneys, bones, eyes, etc. To make conditions ideal for the development of the disease there must be a certain number of bacilli and a favourable soil for them. A good soil is found in a child who has just recovered from whooping cough, measles, etc. People suffering from nervous debility, overstrain, malnutrition, etc., also provide a good soil for tuberculosis. It often happens that a person is infected in childhood and develops tuberculosis of the lungs ; after a time, however, the disease is arrested and the child may be cured. Then later, on reaching maturity, the disease flares up as a result of reinfection. This is what most commonly happens in many cases.

In childhood tubercle bacilli enter the organism. They attack and destroy a small part of the lung (primary focus). The lymphatic glands of the lungs stop the bacilli from spreading while they become infected (primary complex). From that moment until death the bacilli are present in the organism. In this case the reaction to tuberculin is positive. These early changes in the organism can result in a fatal illness or, on the other hand, confer life-long resistance to further infection and immunity to the disease, progress having been arrested at the outset.

The attacked organ can be so diseased that it can no longer function normally. In the early stages of infection the whole organism suffers because the toxin (poison) of the tubercle bacillus circulates in the blood stream.

It is very uncommon for the bacilli to settle and attack several organs at the same time; when this happens, however, it is called acute miliary tuberculosis which is fatal.

The healthy tissues of any attacked organ gradually undergo certain changes and become loaded with bacilli and are finally destroyed.

When the destroyed parts of the lungs are examined bacilli are invariably found. The infection can spread to healthy parts of the lung by the destroyed particles containing bacilli. Other organs can be infected with destroyed pieces of kidney, bone or pus. Destroyed kidney tissue may find its way to the urinary bladder; destroyed parts of the intestines may be found in the excretions, and so on.

All these pathological changes have different effects on the appearance of a consumptive. Some of these effects are obvious and may be noticed by the patient himself while others are less marked and may be observed by the patient's relatives and friends.

FIRST SYMPTOMS OF TUBERCULOSIS OF THE LUNGS

The first symptoms of tuberculosis are loss of weight, with weakness, coughing wi.h expectoration and fever, with sweating. When an ailing person notices that he becomes easily tired and depressed and loses weight he should feel concerned about the cause. If during an ordinary cold there is more coughing than usual or if the cough lasts longer than two weeks it is advisable to consult a doctor without delay. Even a small increase in temperature for a few days between 4 and 5 p.m. may be one of the first symptoms of tuberculosis. Another early symptom is sweating, especially at night. Each and every one of the above symptoms alone might be the first sign of the disease. If a number of these symptoms occur together then the presence of the disease is evident.

It should be noted, however, that many other diseases besides tuberculosis begin with one or more of the above symptoms. We should not therefore think only of tuberculosis. A rising temperature and sweating occur in malaria and in typhus. The temperature varies considerably in individual patients and in different forms of tuberculosis. A dry cough occurs with dry pleurisy, bronchitis, etc.; expectoration with bronchitis, bronchiectasis, abscess of the lung, etc. With each of these diseases there is loss of appetite, loss of weight and weakness. It is obvious that only a doctor can diagnose tuberculosis and modern research has given him many means of distinguishing between one disease and another.

Spitting of blood (hæmoptysis) in apparent good health is also an early symptom of tuberculosis. This symptom often frightens the patient, but is a very useful warning. It is only in some cases that it signifies advanced tuberculosis. At all events *every case of hæmoptysis must be regarded as an indication for radiography.* If the disease is in its initial stage a medical examination of the patient usually reveals very few signs of destruction of lungs, or nothing at all, even if the chest is X-rayed. But even if the patient is assured that nothing is wrong he should be kept under observation. In such cases, if the disease makes progress it would be detected in its early stages and should be properly treated.

As we have already pointed out, even if these early symptoms persist for a long time it should not be taken for granted that advanced consumption is present. If the doctor has made his examination and has said there is nothing to worry about, then the patient should have complete confidence in his knowledge and skill.

We have mentioned only early symptoms which can be noticed by the patient himself, but there are many signs which only a doctor as a result of examination can discover. We need not discuss these, however, as it is essentially a matter for medical men.

To illustrate the symptoms discussed above the following examples are given.

(1) A young girl, guest at a dinner party, often cleared her throat. She told her neighbour, a medical man, that she smoked too much and would have to give it up. The doctor suggested that she should consult a throat specialist to know if there was anything wrong. He further suggested that if the throat specialist was unable to find anything she should come to him for an examination of her chest. The throat specialist found nothing wrong, but the doctor discovered that she had tuberculosis in the upper part of the lungs. She underwent treatment and in time was cured. It should be noted that the only symptom in this case was the patient's constant clearing of the throat.

(2) A young man was working very hard for an examination. His father noticed that the boy was losing weight and was worried about it. He told him to rest, stressing that his health was more important than his studies, but he refused to take care of himself. In time he was called up for a medical examination for military service. He was rejected as he was found to suffer from tuberculosis He had treatment and was cured. It is interesting to observe that he was

not aware of his condition but his father noticed the only striking symptom, the boy's loss of weight.

(3) A middle-aged civil servant was very fond of night life. He was not unduly worried about a hoarse cough and a husky voice with which he had been afflicted for a long time and which he put down to his late nights. After a merry night out he caught a cold. The next day he consulted his doctor who discovered that he was suffering from tuberculosis and that one lung was practically destroyed. The treatment, however, proved successful and he got well. It should be noted that the only symptoms in this case were a hoarse cough and a husky voice.

(4) A mother noticed that her child always had a flushed face and hot hands in the afternoon, and was worried about it. Her friends advised her to give the child some potent medicine which brought down the temperature a little. After two or three weeks the child got worse and the mother consulted a doctor. The child was found to suffer from tuberculosis but was cured after treatment. The only symptom noticed by the mother in this case was a high temperature.

(5) A young woman civil servant was rather concerned about her figure. She wanted to reduce it and in order to do this cut her food down and had massage. She was also working very hard at the time. One day during massage she spat blood. She was found to suffer from tuberculosis which had flared up once again after an early attack had been cured in childhood. In this case the artificial loss of weight together with overwork brought on her illness and spitting of blood was the first symptom.

(6) A young college boy, a good athlete, spat blood which he thought came from his nose. The next day the same thing occurred but in greater quantity. He was examined by a doctor who found tuberculosis. After a long and difficult course of treatment he got well. It should be noted that even a young athlete can have tuberculosis without knowing

it. The only symptom in this case was spitting of blood.

(7) A young girl student often spat blood. She consulted her doctor who found only a small patch on the lung. This went on for two years until she married and in time had a child. After it was born she spat blood again and it was found that she had advanced consumption. She was treated and got well. It is interesting to note that pregnancy caused the tuberculous lesion to become active. The only symptom in this case was spitting of blood.

(8) A young student on a visit to a watering place noticed that he suddenly became tired and perspired. He drank alkaline water and soon after spat blood. He consulted a doctor who told him he had tuberculosis. Treatment in this case was unsuccessful.

DIAGNOSIS

Only a doctor is qualified to make a diagnosis. Medical science has advanced so far that it is easy to diagnose tuberculosis of the lungs or of any other organ and know what form of the disease is present and what stage it has reached. Early diagnosis by modern methods is quickly made to-day. Early diagnosis generally means early cure, and detection is effected by means of X-rays which enable us to photograph the lungs.

In former times a medical examination was mainly dependent on the appearance of the patient and the diagnosis was confirmed by the aid of the stethoscope, laboratory methods, examination of the sputum, and the use of X-rays. To-day the doctor by using all these methods can make a correct diagnosis.

In the early days when examination was made by means of clinical observation and the use of the stethoscope a diagnosis took longer to make than an examination would take now using all the means we have described before. A few years ago it took from half an hour to an hour to obtain an X-ray photograph of one patient but to-day it is possible by means of mass radiography to take a great number of photographs in a very short time.

A single examination with X-rays or microscope by means of which a lesion can be detected is not satisfactory. Finding the bacilli in the sputum of a patient without seeing him or making a complete examination only reveals one thing, that it is dangerous for other people to mix with him. Diagnosis must be the result of examination aided by all methods. It must reveal the form, development, location, activity, etc., of the lesion.

As always, clinical observation and examination together with the patient's own statements are most important in making a complete diagnosis. It must be admitted that by these means alone we cannot always detect tuberculosis in its early stages; this can sometimes be done quickly by X-rays.

A complete and definite diagnosis can only be made by the use of all the methods described.

Sometimes a doctor even after using all the methods of examination cannot be sure whether or not his diagnosis is correct. He must keep the patient under observation for a long time if necessary and then if he finds that the treatment he has been giving brings no results, he must modify it. The patient must be made to understand that it is for his own good that he should be kept under observation and undergo treatment. This is the reason why a private doctor should send his patient to a hospital for observation.

Without a correct diagnosis it is impossible to give correct treatment. It is clear that an unqualified person can neither make a diagnosis nor give correct treatment. In spite of the fact that tuberculosis is caused by only one specific germ the disease appears in many and varied forms and different degrees of activity.

If a hundred people had tuberculosis one could say that they were suffering from a hundred different diseases. In other words there are no diseases but only sick people.

Each case would have to be treated in a different way. *It should be stressed that there is only one certain means of diagnosing tuberculosis: the finding of tubercle bacilli in the sputum. The sputum examination is of primary importance and should be repeated at frequent intervals.*

MASS RADIOGRAPHY AND TOMOGRAPHY

For the diagnosis of pulmonary tuberculosis we have used for many years X-ray photographs as well as the stethoscope, microscope, etc. The use of X-ray photographs has been greatly developed in recent times with two main objects in view. First, to examine in greater detail the nature and development of the lesions, and secondly to examine the greatest number of people in the shortest possible time.

In order to study the lesions in greater detail a new method has been evolved in recent times known as tomography. This is a method of photographing the lungs in such a way that sections of the area photographed can be seen to any required depth. By such a method lesions which would not be clearly defined in an ordinary X-ray photograph are more clearly seen. Sometimes lesions not discovered by ordinary X-ray photographs are detected when tomography is used.

To study a case in detail we must not only make a physical examination and laboratory tests but also take X-ray photographs supplemented by tomography. Taking ordinary photographs requires a great deal of time, and for tomography a high degree of technical skill is essential.

When it was first decided to examine healthy people in rder to detect latent cases of tuberculosis doctors were confronted with the problem of finding the best method to examine the greatest number of people in the shortest time. To make an examination on ordinary lines took a long time and required many doctors and assistants. It was decided that by means of radiography a better and more rapid general impression of each case could be obtained. Only about 20 simple X-ray photographs can be taken in an hour. Much time, equipment and space are needed in order to develop these photographs. It is clearly impossible to handle hundreds of cases daily.

As in ordinary photography large single plates are used, but in mass photography hundreds of small negatives are employed.

In mass radiography it is possible to take and develop hundreds of photographs rapidly. Less space and equipment are needed for this purpose. The result is that hundreds of people can be examined daily by the same personnel as would be required by ordinary X-ray radiography.

By using mass radiography a great number of cases can be photographed daily thus resulting in economy of personnel and equipment. The small negative enlarged for study tells us whether there is danger or not but does not give the details which are obtained by ordinary X-ray photography and to a greater extent by tomography.

When mass radiography is applied at regular intervals (six monthly or yearly) to persons working in the same environment it is called periodic mass radiography. The object being first to eliminate persons who are actual or potential expectorators of the tubercle bacilli, and secondly to detect new cases when the disease is in its early stages even before symptoms manifest themselves.

Examples of mass radiography are the schemes for examination of recruits in the Canadian and United States Armies, in the Australian Imperial Forces, of certain Royal Air Force personnel, and of applicants for employment in various industries and business concerns such as the staff of the Metropolitan Life Insurance Co., New York City, of the Eastman Kodak Co., U.S.A., and of Philips Radio Works, Eindhoven, Holland.

There is no doubt that routine periodic mass radiography is the best method of controlling pulmonary tuberculosis in any community and it is to be hoped that it will be included in all schemes of medical reconstruction throughout Europe.

SHOULD A CONSUMPTIVE BE TOLD THE TRUTH?

Modern research and experience have taught us that if it is taken in time tuberculosis of the lungs can be cured not only in its early stages but even if it is advanced. It is true, however, that sometimes cure is impossible because the patient cannot afford treatment.

What is the point of hiding the truth from a patient when he can be cured? The old idea that tuberculosis is incurable is fast losing ground among the general public and it is to be hoped that in time it will vanish altogether. In the old days to tell a man that he had tuberculosis was like passing a sentence of death.

As we know tuberculosis is in most cases a chronic disease which necessitates prolonged treatment. In time the patient is bound to find out what is wrong with him, and it is far better that he should be informed by a doctor in the proper way than by an unqualified person. He should be told that he is a danger to those living with him and that he can infect them. Those in close contact with him must be warned and the patient must continue to be very careful with his habits and must always use a pocket spittoon. It must be made clear to him that he should take extra care in his contact with children. The sooner things are explained to him the sooner he will take care and so help his doctor to get him well.

It must be admitted, however, that telling a patient the truth is not an easy task. Some patients can be told at once while with others one must be very careful and first prepare the ground. We must, however, inform those in close contact with the patient without delay. The matter must be handled with understanding and sympathy. The main note whenever possible should be optimism. In hopeless cases we must not tell the patient but the situation must be explained to his relatives or friends.

TUBERCULOSIS IS A SOCIAL DISEASE

From some of the examples given in the preceding chapter we have seen how tuberculosis can be spread leading us to think of tuberculosis as a social disease. It is a fact that without tubercle bacilli there can be no tuberculosis. It is also a fact that tubercle bacilli like seeds must find a favourable soil if they are to flourish. It is difficult to say which is of greater importance. the presence of tubercle bacilli or a favourable soil. Both are of equal importance. A favourable soil for tuberculosis is brought about by many causes which in themselves make disease possible and should be carefully considered. These causes are the social causes of tuberculosis and make it a social disease.

When tuberculosis attacks an individual he must give up his work or else he will in time infect his family, the members of which will infect their friends and so the infection will spread throughout the community. This further shows that tuberculosis is a social disease. It is well to bear in mind that the tubercle bacillus is no respecter of persons.

The following two examples also illustrate the fact that tuberculosis is a social disease which can flourish in the homes of the rich and poor alike.

The first case is that of a wealthy family living in excellent surroundings. The mother, however, had tuberculosis. She was very fond of her children and was always kissing them and playing with them. There was no discipline in this household. People went to bed when they liked, danced and drank until the early hours of the morning. The mother's infection was soon caught by all the children and in time they all died.

The second case is that of a poor family whose living conditions were appalling. There were seven people in one room and the father had consumption and infected five children who eventually all died. It is clear that people living in wealthy homes can be attacked by tuberculosis just as well as their more unfortunate fellows whose living conditions are very bad. The poor family lacked food and a decent home, while the rich family lived an undisciplined life.

In some countries tuberculosis is called the people's disease, in others, the white man's scourge.

SOCIAL CAUSES OF TUBERCULOSIS

The causes which predispose to tuberculosis are: insufficient natural immunity (resistance), low standard of living (low earnings), unhygienic homes, under-nourishment, alcoholism, careless living, hard work and long hours, bad working conditions, insecurity of occupation, unemployment, heavy work for women and children, poverty, etc.

A branch of medicine called social medicine deals with these causes, analyses them by studying statistics and records, and makes suitable recommendations.

One of the most important factors not only from an individual point of view but generally and socially is immunity against tuberculosis.

It is not possible to acquire immunity against tuberculosis to the same extent as against smallpox. We can only speak of relative immunity.

In those parts of the world where tuberculosis is unknown the people have no resistance to the disease if it should attack them. If infection is brought to these places the disease in its most acute and fatal form (galloping consumption) spreads far and wide. In those districts where tuberculosis is unknown the standard of living and bad conditions have no effect until the infection is introduced from outside. Whole native populations have been wiped out in Africa when brought into contact with white men at trading stations.

Natural immunity in a nation takes a long time to develop and only comes after a high price in victims has been paid. Immunity develops after a mild infection in childhood repeated many times. These mild infections do not do much harm if social conditions are good. They are more dangerous, however, if social conditions are bad. In the latter case a small number of bacilli will have as great an effect as a large number under good conditions. A large number of bacilli attacking an individual who lives in bad and unhygienic conditions results in grave illness. Civilised people have a certain immunity and their resistance against tuberculosis increases with improvement in social conditions. In Europe this immunity is most marked in the towns, but it is gradually permeating the rural areas. At first, healthy country people came to towns, caught tuberculosis and died; later they acquired immunity both in the towns and in the country. Jews have the greatest immunity to tuberculosis and the lowest mortality. The Jewish race has in the course of time suffered very great losses as a result of tuberculosis but it has acquired a high degree of immunity. This immunity, however, can be lost. In Arabia the Jews lost this naturally acquired immunity and fell victims of the disease as did the natives whose immunity was very low. Another factor in connection with natural immunity is the constitution of the individual.

As we know different animals react in different ways to tuberculosis. Some of them react negatively to massive doses while others react positively to small doses. Human beings react in a similar way.

The most important social factor is community life, not only from the point of view of the dissemination of the disease but from the point of view of the human soil when the infection attacks individuals and groups.

A community without tubercle bacilli will not suffer from tuberculosis whatever the social conditions or lack of resistance may be. But under such conditions the introduction of the bacilli will spread tuberculosis everywhere. In a community where the social conditions are satisfactory and natural immunity is highly developed the tubercle bacilli are not dangerous.

In conclusion the best means of defence against tuberculosis is to raise the general standards of living and education and to make it possible for all to have modern treatment. Under such conditions a mass attack of bacilli will play an important part in the process of natural immunisation.

THE ECONOMIC FACTOR AS A CAUSE OF TUBERCULOSIS

Tuberculous infection makes no distinction between rich and poor, but the results are different. If the infection attacks people in comfortable circumstances where conditions are hygienic it will be difficult for it to gain a foothold. If the infection attacks people in poor circumstances who are unable to build up a strong resistance or where conditions are unhygienic the infection will gain a foothold easily and develop rapidly.

People in comfortable circumstances who are attacked by the disease and can obtain good treatment have a far better chance of being cured. On the other hand poor people who have caught the disease and who are not able to work and earn money, find themselves in a very difficult position to obtain proper treatment. If the sufferer is a family breadwinner, his wife and children will also suffer from want as a result of his illness.

This means that tuberculosis brings untold misery in its train when it attacks a poor family. There are many facts which establish beyond doubt that economic circumstances are a social cause of tuberculosis. If we examine the figures before, during, and after the first World War, we see that tuberculosis was declining before the war, and as a result of the lowering of the standard of living during the war it was increasing, and after the war continued to increase until the standard of living improved.

It is clear that the economic factor plays an important part in the incidence and spread of tuberculosis.

In his well-known work, "The Control of Tuberculosis in England," Dr. G. Gregory Kayne sums up the situation as follows: "In regard to all forms of tuberculosis the improvement of the hygienic and economic living conditions of the working classes must remain a measure 'de premier ordre' for the prevention of tuberculosis."

TUBERCULOSIS AND THE HOME

The home is an important factor in the fight against tuberculosis. A clean, dry, airy and not overcrowded home eliminates one of the main causative factors in the incidence of tuberculosis. This means that it will be more difficult for the disease to gain a foothold. But if it should do so it will not develop so easily, and if on the other hand it does develop it can be fought with a greater measure of success.

An unhygienic home is a good breeding ground for tuberculosis, as it makes it easy for the disease to gain a foothold and spread. A dark, damp, dirty home with insufficient living space lowers the resistance to tuberculosis. Homes of this type are very dangerous for children. In these circumstances isolation of a patient is impossible.

This type of home can be found everywhere, in villages as well as in towns. But the danger is greater in towns owing to congestion. Overcrowded homes are bad enough, but conditions are made worse if a person with tuberculosis has to sleep in the same room as other people, especially children. Things are made even worse if a consumptive has to share a bed with another person. Shortage of houses is also a social cause of tuberculosis.

These facts show that bad home conditions are one of the most important social causes of tuberculosis. The provision of good houses is one of the ways of solving the tuberculosis problem. This has been proved by the decrease of mortality figures, as in Denmark, where good homes are provided for working people and isolation homes for consumptives.

Bad homes have the worst effect upon children. This has been proved by the results of tuberculin reactions. Eighty per cent. of children in their first year examined in bad homes showed a positive tuberculin reaction, while on the other hand, in the good homes only 40% showed a positive reaction.

In conclusion let us stress that bad homes lower resistance, prepare the ground for the disease and facilitate its spread. Good homes on the other hand strengthen resistance and provide a poor soil for the disease.

FOOD AND TUBERCULOSIS

The classical example of tuberculosis due to malnutrition was seen in the first World War. Under-nourishment prepared the ground for tuberculosis. Increasing prices of foodstuffs and decreasing wages also helped the development of tuberculosis. High mortality rates as a result of tuberculosis were reported in prisoners-of-war camps in the last war. Famine reduced resistance and tuberculosis flared up and we were suddenly confronted with an epidemic.

A person who is undernourished or is starving loses strength and falls an easy victim to tuberculosis. This is proved by statistics. Diet is very important. The quality of the food must be good. It must contain all the necessary elements, carbohydrates, fats, proteins and vitamins in balanced proportions. Mineral salts also play an important part in nutrition, and these are contained in vegetables and fruits.

According to an eminent Norwegian authority, Dr. H. J. Ustvedt, the importance of nutrition is so great that it would appear reasonable to regard it as the most vital of all the factors underlying the association between a low standard of life and high incidence of tuberculosis. Dr.

Ustvedt further states that " When attempting to explain the great difference in tuberculosis mortality among different economic levels of the population we are justified in placing nutritional factors in the foremost place."

TUBERCULOSIS AND ALCOHOLISM

Alcohol in small quantities has a certain food value and stimulates the appetite. But alcoholism is harmful in any circumstances and is one of the social causes of tuberculosis. Too much money spent on drink means too little spent on necessities. A family whose bread-winner is a drunkard generally finds its standard of living getting lower and lower. In these conditions it is easy for tuberculosis to gain a foothold or to flare up. It is a well-known fact that owing to a weakened constitution the children of drunkards fall easy victims to consumption. A drunkard who suffers from tuberculosis may also be a dangerous disseminator of infection owing to lack of self-control and discipline.

TUBERCULOSIS AND MODE OF LIFE

A man's way of life is often an important factor as a social cause of tuberculosis.

The life of the countryman is different from that of the town dweller. As more people from the country gradually settled in towns their mode of life changed and an increase in the mortality rate of tuberculosis occurred which was due to new and generally unsatisfactory living conditions.

The new work which they took up in unhygienic and overcrowded factories without the sunlight and fresh air they had always known weakened their resistance to the disease. Another important factor is that country people in general have a lower immunity to tuberculosis than townspeople.

It is clear that there is a close connection between a man's occupation and tuberculosis. It is not so much a matter of how hard a man works but the conditions under which he works. As congestion is not so prevalent in the country tuberculosis does not spread so much among country people as it does among workers in factories, offices, shops, etc.

Records show that figures for tuberculosis differ greatly in different trades and professions.

SOCIAL PATHOLOGY OF TUBERCULOSIS

The incidence of tuberculosis in any country reflects its economic, social and cultural standards.

The mode of life of our ancestors was crude. Civilisation and all it brings in its train is man-made. There is a general desire to make life easier. The products of civilisation, machines, etc., are used to serve mankind and produce more goods for less effort in less time. This applies to town and country alike. In the hands of civilised people machines should provide a living for all but should not be used as a means of exploitation. The products of civilisation are well used if they serve the needs of society as a whole and if they are used to raise the general standard of living and culture. Under such conditions tuberculosis as a social disease would not make much progress and would soon become a rare disease. If, on the other hand, the machines enslave mankind then the advantages of civilisation will benefit only a minority. Every class of society is exposed to tuberculosis. Tuberculosis attacks both those who own and those who work the machines. It follows that where civilisation serves the needs of the people as a whole tuberculosis is on the decrease but where civilisation is made to serve only the favoured few tuberculosis is on the increase. Thus the social pathology of tuberculosis comes about when the majority are slaves to the machines.

In the Middle Ages the majority of people worked on the land and only a small number were tradesmen. For this reason the towns were small and few and town workers were not numerous. Large factories were non-existent. Employers had but one or two assistants. The standard of living of all these people was about the same as was their mode of life. Working hours were short, night work was unknown and there were many holidays. Then came the industrial age. At first the employer continued to work but employed a greater number of assistants. The employer devoted a great deal of time to business management and problems of expansion. Wages were cut down and working hours increased and working conditions became worse.

The advent of the machine age which should have made work easier soon made the worker a slave of the machines, and also a slave of capital. Skilled workers were soon replaced by unskilled ones and often by women and children. Then came unemployment. Labour became cheaper and hours of work longer. Profits had to be paid frequently on capital invested in machinery and factories, and large sums of money were used to buy new equipment and build new establishments. Nobody thought about the health of the workers and those who fell ill or were injured could be easily replaced by other and cheaper workers. Insurance was unknown. Wages were out of all proportion to the work done and the hours of labour. They were just sufficient to maintain life at the lowest level. In time the workers formed Unions in order to get higher wages and better working conditions. They felt the need for these in order to live at a higher and more hygienic level. All this took place in the nineteenth century.

As the general standard of living became worse tuberculosis increased and gradually spread to the villages. It happened in this way. Farm workers came to the towns in search of work and caught the disease and when they returned to the country they spread it. It is easy to understand from the foregoing facts how class divisions and political parties for each class came into being. The formation of workers' parties and Unions was followed by laws setting up better conditions and insurance societies. Wages and working conditions were improved and the Friendly

Society movement flourished. The employer now had to assume some responsibility for the health of the workers. But these measures achieved very little and the incidence of tuberculosis was practically unaffected. It continued to head the mortality list.

The social pathology of tuberculosis arose from the fact that the worker became a mere machine whose fundamental needs were ignored. At that time more was thought of beasts of burden than of the workers. If the former became sick or died it was costly to replace them, but if the latter became incapacitated or died they were easily replaced and the work went on. From this stage the social pathology of tuberculosis developed to the present day when we are only using palliative instead of radical measures against it.

In the last century long working hours were the rule for all trades. Some people worked as long as 20 hours a day. The introduction of the eight-hour day was the first ray of hope.

This was of the utmost importance to women and children. Other improvements including workers' houses near factories etc., came later. Short-sighted employers appear to think that their interests are best served by making their employees work longer hours. This reacts unfavourably on the health of the workers who are not able to put so much energy into their work and in time production suffers, thus reducing their earnings and lowering their standard of living. The worker in many cases has become a robot and his movements are limited to those necessary to perform his task. The work is monotonous and boring. All this has a bad effect on the worker from a psychological point of view. In the past a skilled man made a whole shoe himself, but to-day in the factory he makes only a small part of it. He no longer has the satisfaction of the craftsman in doing the complete job himself. In the old days it took a man many years to become expert at his job, but to-day he can learn to work his machine in a few hours or days.

Tuberculosis developed not only as a result of long hours and monotonous work but also owing to unhygienic conditions in the factories. Unenlightened employers are averse to introducing reforms which are costly and reduce profits.

Hygienic conditions are very important where workers are using materials containing poisons or where it is necessary to have frequent changes of temperature.

As a result of the conditions described above, besides others not mentioned here, tuberculosis has developed in industrial communities and has become largely a workers' disease.

In his admirable work on "Tuberculosis and National Health," Dr. H. Hyslop Thomson states that "Tuberculosis is a disease of crowded communities, the incidence increasing as the social and hygienic standard of such communities deteriorates." Tuberculosis may, therefore, be accepted as an accurate index of the social and health standard of a community.

THE SPREAD OF TUBERCULOSIS

Owing to the social causes under consideration tuberculosis is widespread and under such conditions the tubercle bacillus finds many victims. It is to be greatly deplored that tuberculosis destroys the flower of the nation, its young manhood and womanhood. Tuberculosis is insidious in its approach and it strikes without warning. When the enemy reveals itself it is often too late for defence and the victims succumb easily. Tuberculosis causes a vast number of deaths. But the public is passive in its attitude to the disease. Whenever there is an epidemic of typhus the press and public show a great concern about it, but they are not alive to the danger of tuberculosis.

They are not aware that the danger is always present as an unseen enemy. Every country has lost some of its greatest writers, scientists, statesmen, at the height of their fame, as a result of an attack of tuberculosis. The study of tuberculosis statistics makes if clear that the yearly loss of life to society is enormous.

It is equally clear that the loss to a country in the sphere of economics can be counted in millions of pounds.

By studying the mortality and morbidity figures and the statistics of tuberculin and biological reactions we can have some idea of the spread of tuberculosis. It must be admitted that complete figures for every country are lacking owing to the many difficulties of obtaining them. More accurate figures could be obtained if more care were taken in recording the results of post-mortems, and other examinations, by public authorities in factories, schools, local government offices, etc. Making the notification of tuberculosis compulsory in all countries would also be a great help. Great Britain was one of the first countries to make the notification of tuberculosis compulsory in 1908.

It is possible, however, to obtain a fairly clear picture of the true position from the figures available. From birth a child is exposed to infection. It is safe to say that in urban districts almost everyone has been infected up to the age of 15 to 18, while in rural districts the primary infection might occur up to the age of 25. It is true that there are many parts of the world where people do not come in contact with infection. As an example we may mention Montenegro, the mountainous part of Yugoslavia. Before the first World War tuberculosis was unknown there. It was noticed in the Belgrade high schools that pupils from Montenegro were very susceptible to tuberculosis and died of "galloping consumption." They came from a community free from infection and mixed with people who were suffering from tuberculosis. The Montenegrins were without natural immunity and fell easy victims. Conditions in Montenegro have now changed and owing to contact with the disease the people have acquired natural immunity. This is not the only example that could be given.

It will be remembered that during the first World War large numbers of native troops who came from regions where they had not acquired natural immunity fell victims of the disease in Europe.

Statistics of tuberculin reactions for large towns show that 96% of the population have been infected before reaching manhood.

The figures for children show that 60% up to seven years of age and 70% up to ten years of age give a positive reaction. These figures are much the same everywhere. In the country the percentage is lower but it is on the increase. If it is true to say that overcrowding is responsible for the spread of tuberculosis in the towns it is equally true that a lower standard of education is responsible for the same thing in the country.

It is dangerous to send a consumptive to stay with friends or relatives in the country as this helps to spread the disease.

It is very difficult to obtain accurate figures of the number of people suffering from tuberculosis. But it is thought that the number of cases is five to seven times greater than the number of deaths. It means that for every death due to tuberculosis there are five to seven people suffering from it. It is interesting to study the following figures :

Out of 100,000 persons—

75	died from tuberculosis in	Denmark	(in 1928)
29	,, ,, ,, ,,	U.S.A.	(in 1928)
88	,, ,, ,, ,,	Germany	(in 1929)
96	,, ,, ,, ,,	Great Britain	(in 1929)
134	,, ,, ,, ,,	Italy	(in 1927)
220	,, ,, ,, ,,	France	(in 1939)
280	,, ,, ,, ,,	Yugoslavia	(in 1936)

(*Treatment and Prevention of Tuberculosis—Dr. Milosh Sekulich—Belgrade*, 1935)

Thus, in France, in 1939, 80-90,000 people died from tuberculosis. This is 12% of the general mortality rate. It further shows that at the time, about 500,000 people in France were suffering from tuberculosis.

It has been noticed that the mortality rate is very high among children under one year of age in families where there is tuberculosis. The rate is not so high up to the age of ten. The figures are nearly constant for every age up to ten and then the figures gradually increase and reach the highest level between 20-40, and then decrease.

More than 75% of fatal cases in men occur between the ages of 15-55 and in women between 18-35.

The mortality rate of tuberculosis before the last war was very high but gradually decreased as the level of hygienic, economic and social conditions improved.

During the present war a great increase of tuberculosis has been observed in various European countries. In Norway, where the incidence of tuberculosis had been reduced to a minimum, the number of cases rose from 33,000 in January, 1942, to 55,000 in January, 1943, an increase of nearly 70%.

In France a considerable increase of tuberculosis has been reported. In 1939 the rate was 63 per 1,000; in 1940, 91 per 1,000; in 1941, 75 and in the next year 70 per 1,000.

In Yugoslavia, in 1942, a children's specialist after having inspected a number of institutes for children declared that 60% of all children aged 6-15 were tuberculous. It is estimated that in Yugoslavia there are about one million persons suffering from tuberculosis or being kept under observation.

In Great Britain, where conditions are better than anywhere in Europe, tuberculosis is also on the increase. The latest statistics for Great Britain given below establish this fact.

Year	*England and Wales*	*Scotland*
1938	25,539	3,432
1939	25,623	3,526
1940	28,151	4,003
1941	28,667	4,174
Percentage increase of 1941 over 1938-1939	12.1%	20.0%

(*Parliamentary Debates*—(*Hansard*)—*Vol.* 396, *No.* 28. 10 *Feb.* 1944)

THE FIGHT AGAINST TUBERCULOSIS IN RURAL DISTRICTS

The fight against tuberculosis is being waged in different countries by different means. Even in the same State the methods used in towns differ from those used in the country.

Not one nation can claim to have done everything possible for the country districts, but generally a great deal has been done in urban areas.

In spite of the fact that the peasant lives near to nature and in the fresh air he is not protected against tuberculosis. Tuberculosis penetrates into country districts easily owing to frequent contact between town and country which is one of the main factors in spreading infection. The statistics of recent decades show that tuberculosis in rural districts of different States (Germany, Rumania, Poland, Holland and others including America) is increasing steadily. The position in these countries is getting worse owing to lack of organisation in rural districts, so that we come to the conclusion that the problem of tuberculosis is of greater importance in rural districts than in towns.

It should be noted, first of all, that statistics are largely incomplete, particularly for the rural districts. We will give only one example here concerning Yugoslavia. In 1937, in the most progressive districts, in 60% of fatal cases the diagnosis was uncertain. Tuberculosis is not notifiable by law in Yugoslavia. This means that the same case can be recorded many times or not at all. Statistics must be improved not only in rural districts but also in the towns, even in States where tuberculosis is notifiable. Examples of comprehensive and sound study of this problem in any district are rare. In 1935, the Hygiene Institute in Zagreb (Yugoslavia) undertook the study of the situation in 14 villages in Groatia. They made a complete examination of 10,267 people, using X-rays, and drew up a complete picture of the situation in that particular province. This was made possible by taking a typical village in each district of the province. There were wide differences between one village and another. In some, hygienic, economic and cultural conditions were satisfactory, while in others they left much to be desired.

Out of 10,267 people examined, it was found that 3.84% had tuberculosis and needed treatment. This means that out of 2,371,026 country people about 90,000 were suffering from tuberculosis in Croatia. It was found that the highest mortality occurred among those over 60 (6.55%) and the lowest mortality among those aged 10-19 (2.1%).

A similar situation exists in nearly all the other provinces of Yugoslavia, with the exception of the Drava province, where conditions were definitely better. One of the causes of this unsatisfactory state of affairs is the difference between the social-economic and cultural life in the villages and towns. The educational factor plays an important part. Illiteracy is widespread and the standard of cleanliness is low. A high percentage of people do not use soap for personal or domestic purposes (20% in some districts of Croatia). The economic position is very unsatisfactory because the majority of peasants do not possess the minimum amount of land nor enough cattle. They suffer from malnutrition and they even have to buy bread. A most important factor in the tuberculosis problem is housing. The majority of houses are unhygienic. The question of beds is also important. The average is one bed for two or three people. It may be mentioned that the average peasant is moderate in his habits as can be seen by comparing the amount of tobacco smoked in villages and towns. The townsman smokes seven times as much as the peasant.

The principal cause of tuberculosis morbidity in the country districts is the general prevalence of infection. The soil is favourable because natural resistance is low and where there is malnutrition infectious diseases spread easily.

There is another important fact which Dr. Furlan pointed

out at the meeting of tuberculosis specialists in Belgrade in 1938. The study of tuberculosis in rural districts and in towns brings out an interesting point. The fluctuations of tuberculosis figures for towns and villages are not parallel. The mortality rate among the peasants in the Drava province after the last war rose higher and higher, and exceeded that recorded in urban areas. Recently there has been a parallel decrease in the figures for town and country districts. In the Sava, Littoral and Danube provinces, the mortality rate for townspeople was higher after the first World War than that for country people. At the present time figures are higher for the country than for the towns. In the Drina province the mortality figures are higher for the townspeople in some years and higher for the peasants in other years. In the Vrbas, Zeta, Morava and Vardar provinces, the mortality figures for townspeople are higher than those for the peasants. At the same time the mortality rate for both are on the increase. If we assume that the figures are reasonably accurate there would seem to be a law controlling the incidence of tuberculosis in Yugoslavia. There would seem to be a wave sweeping from the north to the south. This wave first strikes the towns, reaches its peak, and then subsides. This wave reached a peak in the Drava province in the towns and rural districts and is now subsiding. In the Sava Littoral and Danube provinces the wave is subsiding in the towns but rising in the rural districts. In the Drina province the wave has risen to a high peak both in towns and in villages. In the other provinces the wave has not reached as high a peak in the towns as in the country. Dr. Furlan explained this situation as due to the progress of civilisation. When the gifts of civilisation raise the standard of living and improve the economic and cultural conditions, tuberculosis decreases. Under such circumstances an important role is played by naturally acquired resistance which is produced by a relatively small number of bacilli which have been rapidly made innocuous. It is very important to build up a higher standard of living and so produce greater natural resistance against infection and in those cases where infection gains a foothold cure will be made easier in consequence.

To give a general picture of the spread of tuberculosis in Yugoslavia, especially in rural districts, we must make use of the mortality statistics which are available. They are not complete but they will have to serve our purpose for this study.

MORTALITY FROM TUBERCULOSIS IN YUGOSLAVIA

(In proportion to the total population)

1924	(12,701,257)	 38,082
1925	(12,890,908)	 35,806
1926	(13,083,395)	 36,913
1927	(13,278,749)	 38,025
1928	(13,477,028)	 38,580
1929	(13,678,262)	 41,352
1930	(13,882,505)	 36,466
1931	(14,089,792)	 37,946
1932	(14,300,191)	 36,350
1933	(14,513,706)	 33,797
1934	(14,730,427)	 32,392
1935	(14,900,372)	 30,677
1936	(15,173,608)	 29,320
1937	(15,280,000)	 30,266

TOTAL MORTALITY (1924-1937) : 495,972

(*Central Institute of Hygiene—Ministry of Health and Social Welfare—Belgrade*, 1937).

MORTALITY FROM TUBERCULOSIS IN ENGLAND AND WALES

Year	*Estimated mean population*	*Respiratory*	*Other forms*
1924	38,746,000	32,690	8,413
1925	38,820,000	32,382	8,005
1926	39,069,000	30,108	7,417
1927	39,290,000	31,066	7,107
1928	39,482,000	29,799	6,824
1929	39,607,000	31,425	6,565
1930	39,806,000	29,414	6,331
1931	39,988,000	29,658	6,160
1932	40,201,000	27,627	6,031
1933	40,350,000	27,854	5,405
1934	40,467,000	25,682	5,200
1935	40,645,000	24,603	4,598
1936	40,839,000	23,801	4,467
1937	41,031,000	23,970	4,559

TOTAL MORTALITY (1924-1937), Respiratory and other forms : 487,160.

(*National Association for the Prevention of Tuberculosis*).

It will be observed that the mortality from Tuberculosis in Yugoslavia is approximately three times greater than in Great Britain.

PART II

PREVENTION AND TREATMENT

In the prevention and treatment of tuberculosis equal importance must be given to each individual case in the community as a whole. The greatest concern of the State in the field of national health is the prevention and treatment of tuberculosis. This is one of the most important tasks of the Ministry of Health and social policy must be of an integral nature for we cannot separate medical from social work nor the prevention and treatment of the individual from the community. The care of the community cannot be carried out by medical treatment alone but must be strengthened by a policy of general improvement in the standard of living which will build up the people's resistance to disease.

It should be clearly understood that the social treatment of disease cannot be carried out by good intentions alone. Money is needed and must be allocated by a sound organisation. The guiding principle must be to give a better return for labour which will open up possibilities of a healthy life for the community as a whole.

Medical treatment can restore an individual to health but when he returns to normal life and its hardships, malnutrition, over-crowded home, long working hours, etc., he soon falls ill again. The following example illustrates this point. A worker was sent by his Friendly Society to a sanatorium for treatment. His wife and children were left at home without sufficient means of support while he was having the best treatment under ideal conditions. In time the patient got well and was sent back home to his family and to the old bad conditions. Despite making every effort to improve the home conditions, his health suffered and all the good which had been done in the sanatorium was undone and he became ill again. If Friendly Societies are properly organised they should make it possible for people to have treatment and should make adequate provision for patients' families at the same time. In the case just described this was not done.

If the number of tuberculous patients were small it would be possible for private enterprise to provide treatment and all necessary adjuncts. As it is, however, every country has many thousands of sufferers and in such circumstances it is necessary that the State should intervene.

Well-to-do people and employers must bear in mind that in spite of their comfortable living conditions tuberculosis may attack them just as it attacks the poor. When this happens they are prepared to do everything in their power to save themselves and their children. The following example is interesting. A well-to-do family employed a girl as a maid after a long period of unemployment. But when the girl started working consumption developed. She could not afford treatment and she did not tell her employer about her condition. In time she infected the children, and one after the other they died. The maid survived them all but eventually she, too, died. This shows that the prevention and treatment of tuberculosis is a matter for the State and not for the individual. The State must make laws which will render possible the prevention and treatment of tuberculosis in all sections of the community, free of charge. These laws must not be merely paper regulations but must be put into operation.

Institutions for the fight against tuberculosis are very costly to build and are too few in number but this should not discourage us. A vaccine against tuberculosis as effective as that which we have for small-pox would solve the problem. But we should still have the social causes to overcome. This clearly indicates that the fight against tuberculosis is a fight for good health in general. We must not wait for a vaccine but continue to fight against bad social conditions with unrelenting vigour. We can no longer confine our hopes to individual treatment alone but must introduce modern reforms in the field of social medicine.

The fight against tuberculosis as carried out in Germany at one time was conducted on individual lines. This venture involved a great expenditure. Large numbers of sanatoria, hospitals, etc., were built in record time for the treatment of hundreds of thousands of cases. We know what happened when the patients returned to their homes where they came under the influence of all the social causes of tuberculosis.

This failure of the Germans led people to think that treatment in a sanatorium was unsatisfactory. What it did prove, however, was that to fight tuberculosis successfully we must use all the means at our disposal. We must raise the general standard of living and aim at a higher level of education besides treating tuberculosis by medical means. The fight against tuberculosis is the responsibility of the community as a whole and every individual in the community must know all the facts about tuberculosis, how it can be prevented and cured.

HOW TO PREVENT INFECTION

While proper treatment reduces the mortality rate, application of sound defensive measures reduces the morbidity. Primary measures which prevent infection are of the greatest importance. In the defence against tuberculosis we must first of all remove the sources of infection. The most dangerous source is a person with "open" tuberculosis. He must be removed from all contact with healthy people, especially children. Secondly, a strong resistance to tuberculosis must be built up. This means that we must sweep away all social causes which undermine the people's health and prepare a favourable soil for tuberculosis.

There are two ways to prevent infection. (1) Remove the source. (2) Prevent the germs from entering the body.

The best way to avoid infection is to keep away from people with "open" tuberculosis. The only people who should be allowed to come in contact with a consumptive are his doctor and those who nurse him. It should be borne in mind that in some cases a single contact is enough to cause infection. The ideal is complete isolation of the consumptive. This is the first thing to be done but it is also the most costly. All patients with "open" tuberculosis should be sent either to the isolation home if their condition is considered hopeless or to the hospital or sanatorium, if a cure is possible. Well-to-do patients can be isolated in a proper manner at home. We must see to it, however, that all the isolation homes, hospitals and sanatoria we need to take care of all people with "open" tuberculosis are provided. As it is not possible in the present circumstances to isolate everyone with "open" tuberculosis certain precautions should be taken by the public. People who have "bad coughs" should be avoided as much as possible. Spitting should be strictly forbidden and spittoons freely provided. No one with "open" tuberculosis should be without a pocket spittoon. Great care must be taken to clean and disinfect spittoons.

If these precautions are taken a great deal will have been done for the man in the street. It should not be forgotten, however, that various objects can be infected by a patient with "open" tuberculosis and thus the disease may be spread in an indirect way.

PREVENTING TUBERCLE BACILLI ENTERING THE BODY

Tubercle bacilli enter the body by way of the nose (inhaling) and mouth (ingestion). If a healthy man keeps away from people who cough and expectorate he will not be exposed to infection. If he does not handle infected objects germs will not come his way. Unfortunately we are likely to meet people with "open" tuberculosis from time to time. It is therefore wise to adopt the following golden rules. If one has to talk with a person who has a "bad cough" one should always stand well away from him. One should not shake hands with him when it is noticed he puts his hand to his mouth when coughing. A bad habit is to handle food with dirty hands. Children should be told that they must not suck their fingers. Habits are formed in childhood. It is difficult to train children at first but it is worth the trouble. If people are careful about these rules and look after their personal hygiene they are not likely to become infected.

MILK AND PASTEURISATION

Milk can be a common source of tuberculous infection when it is contaminated by the tubercle bacillus which causes tuberculosis of the glands, bones and joints, and cripples a vast number of children year after year. As nearly all bulked milk coming into towns is infected, the danger of consuming raw milk will become evident.

The method of sterilising milk was first advocated by Pasteur and is called after his name. It is defined as raising the temperature of the milk to 140° F., retaining it at that temperature in a suitable holder for 30 minutes, and then rapidly cooling it to 50° F.

The great majority of medical experts are in favour of pasteurisation of all milk in order to eliminate the risk of infection. In a Report by the Special Committee appointed by the People's League of Health to make a survey of tuberculosis of bovine origin in Great Britain in 1932, we find the following conclusion: "In view of the gravity of the problem we are of the opinion that all raw milk for human consumption that does not come from tuberculin tested herds should be adequately pasteurised, or if this is impossible, should be sterilised or boiled."

The total elimination of all tubercular cattle from the herds being impossible it would seem that pasteurisation by law is the only rational solution of the problem under present conditions.

According to Professor G. S. Wilson, " If all milk for human consumption were pasteurised in licensed plants under adequate supervision, it is no exaggeration to say that tuberculosis of bovine origin would be eliminated."

In the 1931 edition of the United States Public Health Service Milk Ordnance compulsory pasteurisation is strongly recommended.

In a great many cities of U.S.A. and Canada pasteurisation is compulsory by law and no milk may be sold to the public unless it has been pasteurised. The result has been a great reduction of tuberculosis of bovine origin in America and in Canada. In Toronto all the milk supply is pasteurised. Thanks to this measure, in Ontario only 9.6% of 500 cases of non-pulmonary tuberculosis among children were due to bovine tubercle bacilli. Furthermore, it is interesting to note that all of these bovine cases came from outside Toronto, none from Toronto itself. In London 90% of the milk supply is pasteurised but in country areas pasteurisation is not carried out to any extent. This is reflected in the mortality figures for since pasteurisation has become general in London the mortality from non-pulmonary tuberculosis has fallen by 76% while in rural areas it has fallen by only 59%. It is clear, therefore, that it is in the country districts that pasteurisation is most needed.

In Yugoslavia there are a few establishments for pasteurisation, especially around Belgrade, but only a small amount of milk consumed in the capital is pasteurised. For the greater part the milk consumed in cities and towns is brought fresh from the country every day by peasants from their own farms. This milk is non-pasteurised. But it must be pointed out that throughout the country people are educated to boil all milk for human consumption and this is usually done as a matter of routine everywhere.

Any post-war scheme of medical reconstruction should

stress the need of a compulsory Pasteurisation Order to cover all urban and more particularly all rural areas. In remote country districts where this is not practicable the people should be instructed to boil all milk for human consumption.

DISINFECTION OF SPUTUM

Spitting should only be permitted into a spittoon which must be disinfected. There are two types of spittoons, pocket and hand. The former must be hermetically sealed and the latter need only have a lid. These should contain a small amount of Lysol. If possible it is better to have four spittoons ; two of the pocket type and two of the hand type. When two are in use the other two can be disinfected. Not only the spittoon but also the sputum must be disinfected. To disinfect the sputum, twice as much disinfectant as the quantity of sputum is required. The sputum must be completely covered with the disinfectant and well mixed and left for four hours and then thrown down the drain. The spittoon should be disinfected by placing it in a bowl and pouring disinfectant over it until it is completely submerged and then left for four hours. It is then washed. Care must be taken that the vessels for disinfecting the spittoon and sputum are not used for any other purpose. In hospitals special rooms and apparatus are used for this purpose.

There are many disinfectants which can be used for this purpose : Lysol, Formol, Kuss solution, etc.

Disinfection by chemicals is a simple procedure but it must be carried out thoroughly. Tubercule bacilli in the sputum are mixed with protein particles. Certain chemical disinfectants such as sublimate of mercury solidify proteins. This means that when the disinfectant comes in contact with the sputum it does not completely penetrate into it. In order to destroy the tubercle bacilli the strongest disinfectants must be used, such as cresol, preparations of chlorine, etc. It is always necessary to mix the sputum with a 50% solution of carbonate of soda in order to liquefy the sputum and thus make it possible for the disinfectant to get through to the bacilli. This is the reason why Kuss solution is most commonly used because it is made up of 8 grms. of potash soap, 10 grms. of sodium carbonate, 40 grms. of formaline in 1 litre of water. This disinfection must last 20 hours.

Disinfection can be carried out by incineration, but for this special spittoons and apparatus are required. Disinfection by boiling the sputum in a solution of 50% bicarbonate of soda for half an hour is another method. Steam can also be used for disinfection. This method is generally used in hospitals for disinfecting a number of spittoons at the same time and requires special equipment. Each spittoon must contain one litre of sodium bicarbonate solution to prevent the sputum solidifying before placing it in the steam apparatus.

DISINFECTION OF TABLE WARE

All tableware used by a consumptive should be boiled for five minutes in a solution of 50% carbonate of soda and then washed.

DISINFECTION OF CLOTHING AND LINEN

A patient should only use linen which can stand boiling in the above solution. For handkerchiefs, bed linen, and table napkins, boiling in water will do if it is not possible to obtain anything better. The best way, however, is to put them in a disinfecting solution for an hour and then boil them in a solution of carbonate of soda.

Silk and other thin garments should be disinfected with formaldehyde. Woollen and cotton garments must be disinfected in a special steam apparatus. Suits, fur coats, hats, etc., must be disinfected with formaldehyde at a temperature of under 60 degrees C. Rubber goods should be boiled for 20 minutes or be placed in formaldehyde at a temperature of 60 degrees C.

DISINFECTION OF BOOKS AND LETTERS

These are placed in a formaldehyde steam apparatus or in a vessel into which a few drops of 40% formaline are poured; the books must be open and the box placed in a warm place. Books and letters of no value are better burned.

DISINFECTION OF ROOMS

During the time that a patient is occupying a room it should be kept as clean as possible. Only the articles which he uses will need to be disinfected. His bedding should be exposed to sunlight from time to time. When the patient gives up his room it must be completely disinfected before it is occupied by another person. The walls should be washed or newly papered and the bedding and furniture should be put in the open and exposed to sunlight from time to time. All hospitals and other medical institutions have a trained staff who can go to private houses on request and carry out the disinfection of rooms, articles, etc. If it is possible to seal the room used by the patient it can be disinfected with formaldehyde and the floors washed with hot disinfectant.

DISINFECTION OF FLOORS

The disinfection of floors (earth or stone) in farms is carried out effectively with lime. Tile or wooden flooring is washed with a solution of 10% carbonate of soda, 5% cresol lather, or 5% lysol. For this work a very hard brush must be used. Dish cloths can be washed with a solution of 2% of sublimate of mercury and then dried. People who do the disinfecting must wear special clothing and shoes which can be cleaned after use.

DISINFECTION OF WALLS

Whitewashed walls are easily disinfected by whitewashing them again. Tiled or distempered walls are washed with a solution of 5% Lysol or a 5% of cresol lather. Tapestry or wooden walls must be sprayed or scrubbed with a solution of 5% of formaline or 2% of sublimate of mercury.

DISINFECTION OF FURNITURE

Wooden objects must be scrubbed with a cloth dipped in a solution of 5% formaline or carbolic acid or 2% of sublimate of mercury and left for 15 minutes to dry. If the objects can be dipped in the above solutions better results will be obtained. Ordinary furniture can be scrubbed with a brush which has been dipped in a solution of 5% formaline or 2% sublimate of mercury.

CALMETTE-GUÉRIN VACCINE (BCG)

Immunisation is an important factor in keeping the people free from infection. It has been more or less successful in fighting small-pox, diphtheria, scarlet fever, typhus, etc. To produce immunisation various vaccines and serums are used. A special vaccine must be used for every different disease. Some of them are also used for treatment.

For many years research workers have been trying to produce a serum or vaccine for the prevention of tuberculosis. Treatment by means of serum or vaccine has up till now been unsuccessful. But in the field of prevention results have been a little more encouraging. Thanks to the discoveries of Professor Calmette and his collaborator Guérin we now have some hope of ultimate success. They have given us the Calmette-Guérin vaccine (BCG). In order to make it clear what this discovery means we will explain the main principles which led them to achieve their results. The German scientist Koch, in 1894, made a very interesting experiment. He injected tubercle bacilli under the skin of a healthy guinea pig. Two weeks later he noticed a swelling and then a wound at the point of injection. At the same time he noticed that the glands nearest to the wound were inflamed. The wound remained open and discharged pus until the death of the animal.

When he repeated the same experiment on a tuberculous guinea pig the wound was small and soon healed. The nearest glands were not inflamed. This scientific observation is called the Koch experiment.

The principle upon which Koch's experiment was based was investigated by the French doctor and scientist Marfan in 1886. He showed that when a person was recovering from an infectious disease he develops in his system a resistance to further infection which cannot produce the disease again. Calmette and Guérin used Koch's experiment in 1907 for making a vaccine against tuberculosis. This vaccine consists of tubercle bacilli of the bovine type of slight virulence which cannot produce the disease. The vaccine produces resistance against infection by ordinary tubercle bacilli. It is interesting to note how they made the bacilli lose their virulence. Calmette and Guerin noticed that when bile is added to the medium for the culture of the bovine type of bacillus, after many transplantations over a long period the bacilli lose their virulence and are incapable of producing tuberculosis of the bovine type. These bacilli cannot cause tuberculosis even in guinea pigs which are the best subjects in which to produce the disease. In this way a strain of bacilli was evolved characterised by such attenuated virulence that it failed to produce the disease experimentally. One of the most important facts about these bacilli is that their virulence never returns. A special strain of tubercle bacilli was evolved by Calmette in a series of experiments extending over a period of 13 years. These experiments showed that inoculation with large doses of these bacilli into healthy animals did no harm. They also noticed that animals could be vaccinated with these bacilli and when inoculated with virulent tubercle bacilli they did not develop the disease. This resistance appeared to last for many years.

In May, 1921, the Calmette-Guérin vaccine was tried for the first time in a healthy child from a tuberculous family.

Since that time many experiments have been made and the results have proved satisfactory. Children who were vaccinated with the Calmette-Guérin vaccine were found to be resistant to many other diseases. Many hundreds of thousands of people were vaccinated in this way in France and in other countries, and the value of this vaccine has received a certain measure of recognition. But nevertheless we have not been able to form a definite opinion in its favour as with small-pox vaccine. It should be made clear that there has been nothing to suggest that this vaccine is harmful. Its efficacy will be established only after many years of research and observation. But at any rate something has been achieved.

The Norwegian Association for the Prevention of Tuberculosis advocates B.C.G. vaccine for healthy persons who are exposed to infection and young tuberculin-negative members of tuberculous families.

The method of carrying out vaccination is of the utmost importance. Not the slightest mistake must be made. There are two important rules. (1) The child must be given the vaccine three times in the first ten days after birth, (2) For the following six to eight weeks it must be kept away from all possible sources of infection.

If the results obtained up to date can be definitely confirmed then the discovery of Calmette-Guérin will be one of the greatest weapons of medical science in its fight against tuberculosis.

PROTECTION OF CHILDREN

As we have already mentioned every child is born free from tuberculous infection. Any infection usually occurs later in childhood. The mortality rate of tuberculosis is very high among infants. We must first of all concentrate on how to protect them. Some children recover from the first infection and acquire a certain amount of immunity but later when they are grown up the disease flares up.

Systematic protection of children must be carried out by all the means at our disposal. As the child gets older and its resistance grows stronger the first infection is not so dangerous and if the disease should develop it will soon yield to treatment and reactivation in the future is unlikely.

Protection must be carried out from birth until maturity. If a child is born into a family where there is a person suffering from tuberculosis it must either be taken away from its home or the consumptive must be segregated. If a child is born from a mother with "open" consumption it must be taken away at once. The mother as the natural protector of the child must fully understand what tuberculosis is. Dispensaries for children are the proper institutions to give all necessary instructions to mothers, nurses, school teachers; all persons who have control of or come in contact with children should know all the facts about tuberculosis. School medical officers examine children from time to time. Day nurseries are set up so that children can be properly looked after while mothers are at work. Nurseries for children of sick mothers are also set up so that mothers with "open" tuberculosis may know that their children are being well looked after and in some cases can be allowed to see them under the strictest supervision. Mothers with "closed" tuberculosis can in certain circumstances be permitted to feed their children but only under medical supervision so that suckling can be stopped if the case should become one of "open" tuberculosis.

In France, an organisation for child protection, known as the Grancher System, was established on very sound lines. Children of consumptive parents were sent to the homes of healthy people in the country where they were under medical supervision. Children stayed in the country in some cases until they were 14 years old and returned home if the parents had been cured. If the children became ill they were at once sent to a sanatorium. Preventoria were set up to keep children under observation. Children's convalescent homes, open-air schools and rest centres were built in order to make the children stronger and more resistant not only to tuberculosis but to disease in general. This organisation aimed not only at fighting tuberculosis scientifically but also at attacking the social causes of the disease and raising the standard of health in general.

Norway and Sweden have in recent years carried out the separation of contacts until about the age of three almost as a routine measure. In England a scheme exists in connection with the London County Council who arrange for children to be boarded out in the country. Furthermore, in a Medical Research Council Report on Tuberculosis in War-time (1942), the Committee "recommends that continued emphasis should be laid on the examination and supervision of all possible child and adult home-contacts by tuberculosis officers, as a substantial contribution to early diagnosis and prevention."

TUBERCULOSIS AND MARRIAGE

Should a consumptive marry? This question is frequently asked. There are two points to be considered. First, will married life make a consumptive worse, and secondly, will it infect his partner and children? A person with "open" tuberculosis must not marry. This must be stressed in the case of a person with "open" and active tuberculosis. A patient marrying in this condition will make things worse for himself and become a source of infection to his family. But if a healthy person knowing all the facts should marry a consumptive, children resulting from this marriage must be taken away at once. A consumptive woman who gives birth to a child weakens her resistance to the disease as a result of the strain of pregnancy and childbirth. An exception can be made only when one of the parents knowing all the facts is able to help the other materially and marries with this end in view. In such a case, however, all precautions must be taken and it is advisable to avoid having children.

A person with inactive tuberculosis, i.e., a person who has recovered from tuberculosis, may marry if his home conditions are good and his occupation not too strenuous. Greater precautions must be taken by a woman who has inactive tuberculosis as child-bearing may cause the disease to flare up. Patients who have completely recovered from tuberculosis may marry as they may be considered healthy.

BUILDING UP RESISTANCE TO TUBERCULOSIS

As we have already pointed out there are two important factors to be considered when tuberculosis appears. First, the action of the bacilli and secondly, the soil in which they take root. A person who has a strong constitution and lives and works under healthy conditions provides a poor soil for bacilli. But a person recovering from an illness or living under bad conditions and working long hours in unhealthy surroundings provides a good soil for the bacilli to take root in.

Building up resistance to tuberculosis means fighting against the social causes of the disease. If we clear slums, eliminate alcoholism, improve working conditions in factories and offices, and do away with malnutrition, we shall have won an important battle in the war against tuberculosis.

EDUCATION AND PROPAGANDA

The public must be made to understand the dangers of tuberculosis by every means at our disposal. Propaganda, by means of radio, films, lectures, books, etc., must be carried out everywhere. It is very important that a start should be made in the schools. Propaganda must be original and optimistic in character, and must be handled by experts if it is to be effective. Tuberculosis should not be made into a bogy.

IMPORTANCE OF SCIENTIFIC WORK

The prevention and treatment of tuberculosis is now based on sound scientific lines and on the most up-to-date methods which are applied to suit local conditions. All new discoveries and methods are tested as soon as they are made known to doctors, social workers, and others who are fighting against tuberculosis, and when a new method is tested and proved to be scientifically sound it is put into general use.

INDIVIDUAL TREATMENT OF TUBERCULOSIS OF THE LUNGS

A few decades ago tuberculosis was considered incurable. Consumptives were left to their fate and had to depend upon any natural resistance they might have. At that time the mortality rate was at least 50% higher than it is to-day. Towards the end of the last century and especially at the beginning of this century active individual treatment was applied which resulted in a great reduction in the mortality rate. The introduction of a strict hygienic and dietetic régime and later artificial pneumothorax made it possible to say that tuberculosis of the lungs is curable.

The treatment of tuberculosis of the lungs varies in duration. In early forms it may take a few months while in advanced forms a few years.

AIMS AND METHODS OF INDIVIDUAL TREATMENT

The aim of modern methods of individual treatment of tuberculosis is the complete disappearance of lesions (complete cure) and the recovery of the patient so that he will be able to resume his occupation and live a normal life. Modern methods of treatment aim at eliminating all symptoms until a cure is finally achieved and no signs of the disease are manifest.

A parallel aim of modern methods is to destroy the tubercle bacilli or at least to decrease their virulence and remove their poisonous products, and in this way restore the patient to health.

It is not always possible to obtain a complete cure, i.e., the elimination of the bacilli and the disappearance of lesions. We have to be satisfied with a practical cure, i.e., inactive lesions and less active and virulent bacilli. Even in these conditions a patient can return to his work and family life but must take more care than a patient who has been completely cured.

Modern methods can reach their goal by action on local lesions and on the general condition of the patient. They act on the local lesions by reducing their activity partly or completely, and on the general condition by increasing the resistance and making the system stronger.

As already stated we have not yet found a specific remedy to cure tuberculosis of the lungs. But there are a few remedies such as gold preparations which decrease the virulence of the bacilli. These, together with a hygienic and dietetic régime, increase the resistance to the disease.

The hygienic and dietetic regime acts both on the local lesions and on the general condition of the patient. It has the effect of resting the lungs but not interfering with their functions. Its main object is to act on the general condition of the patient. The most important method acting directly on the local lesions is artificial pneumothorax. Its effect is to stop the lung functioning. When this is done the toxins of the tubercle bacilli can no longer circulate in the system and this renders the action of the hygienic and dietetic régime more effective.

It is for the doctor to decide which method must be used. But in order to satisfy the patient we must explain what is being done and seek his co-operation. A patient must have complete confidence in his doctor's knowledge and skill.

METHODS OF INDIVIDUAL TREATMENT

Methods of individual treatment are divided into three groups. The first group comprises the hygienic and dietetic regime. These measures are more or less compulsory for all patients. The second group consists of special methods which can be used separately or in combination but are always supplemented by the hygienic and dietetic regime. The third group consists of medicines to relieve symptoms, e.g., cough, temperature, anæmia, etc.

1. Hygienic and dietetic regime.
   - (a) Fresh air
   - (b) Rest
   - (c) Food
2. Special methods.
   - (a) Immunotherapy
     - (i) Active immunisation
     - (ii) Treatment by Vaccines
   - (b) Treatment by chemical substances
   - (c) Collapse–therapy
   - (d) Closed suction drainage
3. Medicinal Treatment

HYGIENIC AND DIETETIC REGIME

The hygienic and dietetic régime is used in every case and helps to build up resistance and make the system stronger. This régime alone can in some cases cure pulmonary tuberculosis. It has been used for the last decades together with other special measures. This régime is prescribed for every patient under varying conditions : in the home of the patient, in hospitals and sanatoria, in urban as well as in rural districts. This régime in the treatment of a well-to-do patient achieves far better results than in the case of a poor patient whose home conditions have a detrimental effect on the treatment. A farmer can have comfortable quarters and good food at home or in a sanatorium while a poor patient has, in most cases, to sleep in an unhealthy atmosphere and has to eat food of inferior quality. Treating a tuberculous patient in a hospital or sanatorium is a far different matter from treating him at home. Good results are obtained under all conditions, however, if the treatment is applied with skill and the doctor's directions are strictly carried out by the patient. The results are generally more satisfactory if the patient is removed to a sanatorium where he is free from home cares and worries, and where he is learning the hygienic and dietetic way of life and forming habits which he will find most helpful when he returns to his home and work.

FRESH AIR

Many years of experience have taught us the great value of fresh air in the fight against tuberculosis. It goes without saying that the air of the country is purer than that of the towns. Clean air does not irritate the lungs and regulates the body's temperature through the skin. Constant fresh air makes a patient indifferent to changes of temperature so that he will not easily catch cold. In this connection an important part is played by clothing. The patient must be protected from strong sunlight and strong winds. His room must face south or south-east. It must be possible to have plenty of ventilation. If the room has a balcony where the patient can lie down, so much the better. In forming new habits it is more difficult to train a person of 50 or so than a young patient. The patient must spend day and night in the fresh air. In very severe weather, however, he may have the windows closed if the room is well ventilated. All rooms used by the patient must be spacious, easy to clean and well ventilated. The rooms must not be overheated. The patient soon gets used to the routine and acquires habits which he will not break on returning to his former life. The patient gradually converts his family to his new way of life. In time, fog or intense cold have no effect upon him. Even in midwinter he will not need a hot-water bottle if he is given three good blankets. It is advisable, however, to warm his room a little before he gets up in the morning. The good results which have been obtained by fresh air treatment have led to the theory and practice of climatic treatment.

CLIMATIC TREATMENT

The Egyptians and the Romans recognised the value of climatic treatment. Based on scientific research of climatic factors it was first applied in the middle of the nineteenth century and only a few decades ago came into universal use. In 1840, George Bodington founded the first sanatorium in England and practised the "open-air" treatment of pulmonary tuberculosis. Bodington was without honour in his own country, but in Germany his teaching was accepted and sanatoria were built, notably by Brehmer. Long ago a hygienic and dietetic régime was successfully adopted in sunny Egypt as well as in northern latitudes. But as medical science progressed and the treatment of tuberculosis by means of artificial pneumothorax and other measures gained ascendancy, the importance of the hygienic and dietetic régime became less apparent.

Climate depends upon geological and meteorological conditions. Three conditions are of great importance: geographical position, altitude above sea level, and nearness to the sea. The climate is determined by altitude and temperature. Climate has either a good or bad effect on the system. Changes in atmospheric pressure, differences in night and day temperatures, differences in temperature in the sun and shade, winter and summer temperatures, all have an important effect on the patient. Sunshine influences the circulation of the blood, respiration and other bodily functions so that inactive lesions may become active. It is important to know the proportion of moisture in the atmosphere, quality of the air, rainfall, number of windy days and other data concerning the district.

Climate is classified into two types. (1) Above sea level. (2) Seaside. The climate above sea level is subdivided into above and below 400 metres, the latter is called continental climate and is characterised by a very hot summer and cold winter, and great differences between day and night temperature, as well as by dry air. This climate is suitable for tuberculous patients but only in parts which are far away from towns and their polluted atmosphere and are sheltered from winds, and if changes in temperature are not too great.

The climate above 400 metres is ideal. The variations are limited and for this reason it is particularly suitable for tuberculous patients and can be well tolerated by them. The advantages of this type of climate are smaller atmospheric pressure, lower temperature, pure and dry air, moderate wind, strong sunlight. This climate is excellent for tuberculous patients and has the effect of causing active lesions to become inactive. Some people can tolerate this climate all the year round but others only during certain months of the year. The best seasons are summer and winter rather than autumn and spring.

(2) *Seaside climate.* This type of climate is well known for its invigorating air which is rich in oxygen and other elements, high atmospheric pressure, even temperature and moist atmosphere, long hours of sunshine, pure air and changeable winds. The seaside climate is most useful in the treatment of surgical tuberculosis (tuberculosis of the bones and glands, etc.) and in exceptional cases of pulmonary tuberculosis.

REST

The second important factor in the hygienic and dietetic régime is rest, i.e., complete physical and mental relaxation. Brehmer was the first to advocate complete and prolonged rest and his pupil Dettweiler perfected the method. Work and muscular activity increase the circulation of the blood and cause energy to be burnt up at a greater rate which is harmful to the tuberculous patient.

Patients suffering from acute forms of pulmonary tuberculosis and chronic cases who are getting worse must have complete rest. An increase of temperature (normal 98.4 degrees F.) and loss of weight are clinical symptoms which show not only the gravity of the disease but the need for complete or relative rest. For complete rest the patient must be in bed, in a well-lit and ventilated room. He must not read, talk or get up. Complete rest is given to every new arrival at the sanatorium until an examination is made. When signs of activity have subsided the temperature is reduced to normal and there is no further loss of weight. The patient is then put on relative or partial rest. He is sent to rest on the balcony in the open air. The signs of activity show whether or not the patient should stay for shorter or longer periods in the open air. Even if things are going very well the patient must from time to time have complete rest. Patients are generally advised to rest between the hours of 2—5 in the afternoon, i.e., the time of the day when the temperature is highest. The best way to secure complete rest is to make the patient lie in the recumbent position.

EXERCISE AND RECREATION

When the temperature is down to normal a little exercise is allowed. Later the patient takes exercise and rest alternately. Strong sun, wind, rain and fog must be avoided. It must always be borne in mind that tuberculosis of the lungs requires rest and the amount of exercise allowed must be governed by the appearance of activity. If signs of activity appear exercise must be stopped at once. When activity dies down exercise can be resumed. Temperature, pulse, respiration and weight must be watched at all times.

Walking should be restricted to about a quarter of a mile along a flat road, and back; half a mile in all. The patient should take about 15 minutes to cover the whole distance. He should not stop on the way and his reactions should be tested immediately he returns. The patient must be lying down when the tests are carried out. The temperature and pulse rate are taken immediately on his return from the walk, then five minutes, half an hour, and an hour later. The rate of fall between these times helps the doctor to decide whether or not the patient may continue to take exercise or go on resting.

It must be clearly understood that this walk is designed for test purposes and it must be a rule that a patient at all other times should rest before and not after he feels tired.

After this "flat test" the patient can try a little uphill walk. All this is based upon a determined plan as is the whole routine of the sanatorium.

Recreation is not allowed when signs of activity are present. When, however, the patient is allowed to take exercise he may then visit the recreation room. He may indulge in quiet games such as chess, cards, etc., or read books, and take part in conversation. He may also attend concerts and film shows in the sanatorium. All sanatoria have well-equipped rooms for recreation.

As the patient's physical condition improves he is allowed to spend more time in the recreation rooms with his fellow patients. His mental outlook improves as time goes on and he finds it much easier to bear his long stay in the sanatorium as his attitude to life becomes more cheerful. Great attention must be paid to the psychological aspect of the treatment as the patient will have to stay in the sanatorium for at least three months and in most cases much longer. After the patient has successfully come through his walking tests he gradually goes on to a point where he may do some kind of work.

SPORTS AND WORK

Sports may be allowed but not those which make a great call on the muscles of the chest and arms. Tennis, football and cycling are forbidden. The best sport for a tuberculous patient is golf because in the course of the game the player walks on the level with occasional small climbs, he stops for a while and then goes on, and all the time he is in the fresh air in pleasant surroundings. Winter sports are not so suitable. Patients who can skate may do so gently.

After a patient has been allowed to indulge in sports and exercises he may then be allowed to work. This is very important for the manual worker. From a psychological point of view it is interesting to note that from the moment a patient is allowed to start work his whole outlook improves and he begins to get well much more quickly. Light work can be started in the sanatorium and is then continued in the colony. Sometimes the patient is afraid to start work but every effort must be made to remove this fear. It may happen that a patient has become lazy and does not want to work. Some way of persuading him must be found.

The work done in the sanatorium or colony must be under skilled supervision. This work is called occupational therapy. Doctors and all those concerned with the patient's well-being must see to it that he does not lose the will to work and as a result become depressed.

HYDROTHERAPY

When a patient has been taking exercise or has been doing light work for a certain time he may be treated by hydrotherapy. Hydrotherapy must be applied gradually. Stimulation by mechanical means refreshes the patient and makes him stronger. This results in the patient becoming resistant to cold. With slight and inactive cases hydrotherapy can be started at once. Before hydrotherapy treatment the skin is rubbed with a dry towel followed by friction with alcohol. Hydrotherapy is applied with warm water which is gradually made colder. Anything that tires or excites a patient must be avoided.

HELIOTHERAPY

Heliotherapy is treatment by different kinds of natural and artificial rays. Air bathing consists in walking in the open air in bathing costume and sun-bathing in lying undressed in the sunshine. In winter use is made of ultra-violet rays. All these methods by stimulating the skin regulate the heat of the body, quicken metabolism, respiration and the work of the heart. Heliotherapy is recommended for cases in which signs of activity are either disappearing or have disappeared. On the other hand heliotherapy is contra-indicated in cases with active or destructive forms of tuberculosis. Sun-bathing is forbidden for all patients who spit blood. Treatment by ultra-violet rays is good for patients with anæmia, general weakness, and for debilitated children. Such methods must be applied gradually, and carefully at first for a few minutes and then longer. Heliotherapy is more important for surgical tuberculosis and gives the best results if used at high altitudes or at the seaside.

DIET

Food is the third important factor in the hygienic and dietetic régime. It must nourish and build up the body and replace used energy. Good food in proper quantity strengthens the body and makes it resistant. It is very important in cases of tuberculosis of the lungs.

A Norwegian authority, Ansgard Overland, has shown by his work in Norwegian military schools that a quantitative improvement in diet can lead to a fall in tuberculosis mortality.

In order to nourish the body and replace lost engergy a patient must eat more food than a healthy person of the same height, weight and age. The average worker needs

3,000 calories. To induce a patient to eat the required quantity of food it should be made palatable. Weight must be recorded once a week. The weight of a patient should be built up to five or six pounds above normal.

A normal day's food containing 3,000 calories must contain 500 grams of carbohydrates, 70 grams of fat and 100 grams of protein, as well as vitamins and mineral salts. For every kilogram of body weight for a healthy person the allowance is 32 calories and for a tuberculous patient 45 calories.

Proteins are body building agents. They are very important in the treatment of tuberculosis for maintaining strength and replacing lost tissues. Experiments have shown that if guinea pigs are fed on proteins they do not lose weight and are more resistant to tuberculosis than when fed on carbohydrates and fat only. The best way of obtaining the greatest nourishment from meat is to serve it underdone or raw when it is easily digested. Unfortunately most people cannot eat raw or underdone meat. For this reason meat juice is given as a medicine with very good results. In addition to meat, carbohydrates and fat must be given in proper proportions. Calcium salts are also useful for tuberculous patients. Some authorities favour a certain amount of alcohol while others do not.

The question of diet in pulmonary tuberculosis is complicated by the difficult problem of anorexia (lack of appetite) which is often a prominent symptom requiring constant attention.

As we have said both quantity and quality of food are equally important but diet must be varied. Different kinds of meat, cheese, milk products and vegetables must be served. A patient must have three or four good meals a day. Breakfast consisting of milk or tea, bread, eggs, cheese or bacon. Lunch: soup, fish, meat, vegetables, cheese and fruit. It is a good plan to add 100 gr. of meat juice to the soup. Dinner is on the same lines as lunch.

Light snacks of milk and toast may be given at 10 a.m. and 5 p.m. Such is the régime for the majority of tuberculous patients. If, however, the patient suffers from another disease, or in case of complications, the routine must be modified.

A greater amount of protein is sometimes given in certain cases. This is known as the protein diet which consists of an additional 20-25 grams of protein in the form of eggs, meat or raw meat juice. Sometimes the diet is made richer in carbohydrates by increasing the amounts of sugar, bread, macaroni, etc., but care must be taken not to give an excess of carbohydrates which may give rise to fermentation in the intestines and produce inflammation. Fat in the diet may also be increased from 1 to 2 grams of butter, meat-fat, etc., per kilogram of body weight. An extra amount of fat should be given with caution as it may derange the functions of the stomach and liver.

A salt-restricted diet as recommended by Gerson-Hermansdorfer and Sauerbruch has been tried in the treatment of all forms of tuberculosis. Salt is strictly forbidden. Mineralogen is given to replace ordinary salt. The results of this particular treatment have not come up to expectations. In some cases the results were unsuccessful and the patients lost appetite and weight owing to disturbed gastric functions. While unsatisfactory results were recorded in the treatment of pulmonary tuberculosis with the salt-less diet, better results have been obtained in the treatment of tuberculosis of the skin and surgical tuberculosis.

The ordinary régime can be often improved by the addition of milk and its products. Some sanatoria give two or three pints of milk extra. Some patients can tolerate this while others are liable to suffer from diarrhœa or constipation. Generally half a pint extra with meals is quite safe in most cases.

Such are the dietetic principles for all forms of pulmonary tuberculosis especially in chronic cases. It is clear that in acute forms of tuberculosis or during the acute fluctuations of chronic tuberculosis the diet must be light and restricted to milk, eggs, stewed fruit, raw meat juice, etc.

A pregnant woman suffering from tuberculosis must eat well but her diet must be principally made up of carbohydrates with a small quantity of eggs and meat. Patients with heart trouble must be given very light diet while those suffering from diabetes are treated by special methods. Some tuberculous patients may suffer from obesity and in such cases prescribing a diet is very difficult for obesity must be treated as if tuberculosis were not present. Disease of the kidneys demands a special diet of carbohydrates and fat. In any case special attention must be paid to the digestive organs. If too much food is given the first symptom is diarrhœa. This means that too much is as bad as too little. It is not advisable to make a patient too fat.

Digestive troubles may be either functional or organic.

Functional disorders may be caused by glandular disturbances or by bowel fermentation or as a result of nervous strain. This may occur from time to time in every form of pulmonary tuberculosis. As pointed out before, these troubles may be the first signs of pulmonary tuberculosis. The cause may be attributed to the toxins of the tubercle bacilli or the destroyed lung tissue. The stomach and other digestive organs in these cases become sensitive, and this leads to psychological and physiological troubles. A mistake in applying the régime or in giving medicines may cause these troubles. Sometimes stomach troubles reveal themselves only by loss of appetite, caused by inflammation of the throat or gums. Then a special diet is indicated. Besides loss of appetite which is a symptom of stomach trouble there may also be gastric discomfort. These disturbances may result from too much dieting, bad teeth, excessive medication or fatigue.

Organic troubles give rise to similar symptoms but the causes are different. Organic lesions in the gastro-intestinal system at first manifest themselves by diarrhœa and other dyspeptic symptoms and later typical signs of organic disease appear which can only be diagnosed by a doctor. Such cases call for special treatment. Gastro-intestinal complications are a great hindrance in the treatment of pulmonary tuberculosis. A most important factor in treating tuberculosis is that the patient should have a sound digestion so that he may be properly nourished.

SPECIAL METHODS IN THE TREATMENT OF PULMONARY TUBERCULOSIS

IMMUNOTHERAPY

This treatment is called specific treatment and makes use of dead germs or their products. This method has given good results in the treatment of many infectious diseases. The same principle has been applied in the treatment of tuberculosis.

Immunisation may be either active or passive. In active immunisation dead germs (treatment by bacteria) are used, or their products (tuberculin therapy) while in the passive form, serum from a convalescent (serotherapy) is used.

This immunotherapy is used in certain forms of pulmonary tuberculosis together with the hygienic and dietetic régime. Complete specific action does not occur in the case of tuberculosis as in other infections. Immunotherapy in tuberculosis strengthens the resistance of the organism.

When tuberculin was first used for active immunisation it was found to be harmful, because it was unskilfully applied. Correct dosage was not determined and it was applied to unsuitable cases. At the present time medical opinion on the value of tuberculin is divided. Some doctors use tuberculin successfully in small and correct doses and only in certain forms of tuberculosis while others never make use of it. Skilfully and carefully used in selected cases good results are obtained when a certain natural immunity and resistance exist in the organism.

CHEMO-THERAPY

Treatment by chemical substances with the object of destroying the tubercle bacilli in the human organism has not been successful. All chemical agents which have destroyed the tubercle bacilli in the laboratory have failed to do so in the living body. A great number of chemical substances has been tried : metals, arsenic, calcium, dyes, etc. The most recent are gold preparations. While all these substances have no specific action nevertheless some beneficial results have been reported. Gold preparations do not act specifically but appear to strengthen the resistance of the system to the disease. These preparations together with other forms of treatment may prove useful.

COLLAPSE THERAPY

This treatment is based on the principle that if the lung is rested by decreasing its function, or putting it out of action by collapsing it, the lesions will heal. Collapsing the lung means preventing air entering it during breathing. The lung will then be deflated which means it will have collapsed. In respiration the lung is inflated like a balloon. If the air is drawn out of the balloon it is deflated or collapsed.

The lungs are situated in the chest, one on each side of the heart. A membrane covers the lung and enfolds it in the interior of the chest like a closed sack. This membrane is called the pleura. Artificial collapse of the lung can be brought about when the air from outside is allowed to enter the pleura. This is possible because the air pressure in the pleura is negative and if a puncture is made in it the air from outside rushes in and the lung collapses. Collapsing the lung can also be effected not only by allowing air to enter the pleura but by decreasing the size of the wall of the chest.

COLLAPSE BY OUTSIDE PRESSURE

The most simple way of collapsing the lung is brought about by outside pressure. The patient lies with his affected side on a hard pillow. The weight of the body pressing on the pillow prevents the ribs moving and thus the diseased lung does less work. The patient at first is made to lie in this position for one or two hours a day. The period is increased until he can stand it day and night. This treatment must be used on and off for many months or even a year, together with the hygienic and dietetic régime, and is suitable only in certain forms of tuberculosis. It is the simplest method of producing artificial collapse but it is seldom used.

ARTIFICIAL PNEUMOTHORAX (A.P.)

As stated before collapse of the lung takes place when outside air is admitted into the pleura. Sudden inlet of air is dangerous as the patient cannot stand it and the consequences may prove fatal. If air enters directly from outside the pleura may become infected. For the purpose of performing artificial pneumothorax a special apparatus is used. By means of this apparatus the air which is first sterilised enters the pleura slowly and without any pressure. It is very easily absorbed by the pleura and the operation is repeated at intervals. Complete collapse of the lung must be produced gradually and then maintained.

Artificial pneumothorax can be applied to both lungs.

The chief indication for pneumothorax treatment is the presence of tuberculous cavitation.

Results of artificial pneumothorax depend on whether or not the collapse is complete, on the degree of activity and extent of the lesion and on the presence of adhesions and the possibility of removing them.

Results have shown that artificial pneumothorax is the most effective method of treating tuberculosis of the lungs. The tuberculous lesions disappear and are transformed into scars. When this has taken place artificial pneumothorax is discontinued and in time the lung resumes its original position and function.

The duration of pneumothorax treatment depends upon the particular features of the case. In Ustvedt's opinion it should rarely be less than two years, even when the patient becomes symptom-free and remains well.

OLEOTHORAX

If in spite of repeated artificial pneumothorax a cavity is not closed the operation of oleothorax is performed. While artificial pneumothorax is a method of collapse therapy oleothorax is a method of collapse-compression therapy. In using this method more care must be taken than when performing artificial pneumothorax so as to avoid complications. Certain antiseptics are used such as gomenol, medicated olive oil, paraffin or special preparations such as iodipin. These oils are not only used for producing collapse and compression but for their antiseptic effects against possible exudations consisting of varying amounts of pus. Thus the action of oleothorax is threefold : first, it collapses the lung ; secondly, it compresses it ; thirdly, it disinfects the pleura.

This method has been employed in place of artificial pneumothorax for many reasons. First of all in order to avoid the frequent admission of air. Oil is used because it is absorbed more slowly. Secondly, by using antiseptic oils there is a possibility of avoiding complications such as exudations already referred to. Thirdly, to prevent the formation of adhesions in the pleura. Oleothorax treatment is not free from complications and the results are difficult to assess.

PNEUMOLYSIS OR CUTTING PLEURAL ADHESIONS

One of the causes of unsatisfactory results of artificial pneumothorax is the presence of adhesions in the pleura which prevent the lung from completely collapsing. It is interesting to note that adhesions are generally found in that part of the lung in which there are lesions and it must be collapsed in order to obtain the best results. These adhesions are caused by inflammation around the lesion.

Incomplete artificial pneumothorax becomes complete when adhesions are cut. This involves a special operation called pneumolysis, which means separation of the lungs from the pleura. The results of artificial pneumothorax are more satisfactory when successful pneumolysis is performed.

This operation was first performed by Jacobeus of Stockholm who, in 1913, published his results obtained in fifteen cases. In the early days there were some complications, even fatal ones, but gradually the technique was improved and the operation can now be described as a simple one. Complications are still possible but can be easily dealt with.

PHRENIC NERVE OPERATIONS

The diaphragm which separates the abdomen from the chest plays a most important part in the function of breathing. The diaphragm is a large flat muscle controlled by two nerves (Phrenic nerves) situated on each side of the neck and passing down through the chest. If one of these nerves is cut at the level of the neck the result is paralysis of the diaphragm on the same side. The paralysed part of the diaphragm then rises because of the negative pressure above. This causes the lung on that side of the chest to collapse partially and slows down its functions. The operation of cutting the phrenic nerve is called phrenicotomy and is no longer used because the nerve is regenerated after a time and the diaphragm begins to function again. An operation for which good results are claimed is avulsion of the phrenic nerve which paralyses the diaphragm and allows it to ascend. The nerve is cut at the level of the neck and the lower end is dragged up by gentle traction, thus effecting complete removal of the nerve.

This method of treatment is used when artificial pneumothorax cannot be performed. Permanent paralysis of the diaphragm causes partial collapse of the lung while decreasing its functions and thereby putting the lung at rest. By such operative measures the lesions of some forms of pulmonary tuberculosis can be made inactive.

PHRENIC ALCOHOLISATION

Instead of cutting the nerve to paralyse the diaphragm it has been recommended that the nerve should be injected with alcohol. About 4-6 c.c. of 90% solution of alcohol is injected at the site where the cut is generally made in phrenicotomy. In this way the diaphragm is paralysed for a period which may extend to a year. But before that period elapses the nerve may be restored and the diaphragm may start functioning again. This operation is useful if it is expected that the lesions are likely to heal within a year. If there is reason to believe that the lesions may heal in a shorter time the injection is made with 70-80% alcohol solution instead of a 90% solution. This method, like several others, is only used in the treatment of certain forms of tuberculosis.

THORACOPLASTY

If resting the lung by means of paralysis of the diaphragm does not give satisfactiory results resort may be had to thoracoplasty. This operation consists in cutting out ribs from the side where it is proposed to make the collapse. This method is even older than artificial pneumothorax and was originally devised by Ostlander, but the operation was suggested by Brauer and introduced with a certain amount of success by the well-known Berlin surgeon, Sauerbruch. A full account of this method of treatment has been given by the American surgeon Alexander in 1925 who described the results in 1,159 cases treated by thoracoplasty.

Thoracoplasty has the same end in view as artificial pneumothorax. By means of it we can treat the complications arising from artificial pneumothorax. In spite of the fact that thoracoplasty is an older operation than artificial pneumothorax it did not come into general use until the good results of artificial pneumothorax and other methods had been obtained. To-day we can make an examination and decide whether to perform artificial pneumothorax or thoracoplasty as the first step. In some cases thoracoplasty may complete a cure where artificial pneumothorax or other methods have only been partially successful.

EXTRAPLEURAL PNEUMOLYSIS

Thoracoplasty deforms the chest. For this reason it is not popular with patients. Doctors therefore returned to an old method of operation called Extra-pleural Pneumolysis. This operation was used when collapse was necessary but could not be obtained by pneumothorax and phrenicectomy was not successful. This operation (Extra-pleural Pneumolysis) was designed to separate the lung together with both membranes of the pleura from the wall of the chest in order to bring about the collapse of the lung. To perform this operation resection of one rib is necessary. This operation was first performed in 1891 by Tuffier, a Paris surgeon. He only separated the apex and published the results of his method in 1895 which he called "Apicolysis." To maintain the collapse it is necessary to inject some substance into the lung and pleura. Different surgeons use different substances: paraffin, sponge, muscle, free fat, etc. Owing to various complications such as the appearance of exudate, pus, etc., air is more frequently used (Extra-pleural Pneumothorax). The air has to be let in in the same way as in pleural pneumothorax.

CLOSED SUCTION DRAINAGE

Attempts to close a cavity by collapse measures may prove unsuccessful and to achieve the desired result the operation of closed suction drainage is performed in accordance with the principles laid down by Monaldi, Price Thomas, and others. This operation is often supplemented by thoracoplasty and phrenicectomy.

COMBINATION OF METHODS

If it is not possible to obtain the desired results with a certain method then a combination of methods may be used.

As we have indicated, collapse of the lungs may be effected by a variety of methods which can be combined. Which method should be used for any particular case is a matter for the operator to decide, and the information given above is intended to help the patient to understand the nature of his illness and to trust his physician.

TREATMENT OF TUBERCULOUS DIABETICS

A tuberculous patient may also suffer from other diseases. Treatment in such cases is more difficult. As an example we will take diabetes because it best illustrates the difficulties of the doctor and his patient. While in the case of an ordinary tuberculous patient we are able to recommend a wide range of foods, in the case of the diabetic we must lay stress on a very strict diet.

The treatment of a diabetic with tuberculosis must be devised to treat diabetes and tuberculosis simultaneously. Diabetes is a disease characterised by a disturbance of carbohydrate, fat, and protein metabolism. The cause of this disease is the faulty functioning of certain endocrine glands, particularly the pancreas which secretes insulin.

Owing to faulty assimilation of food it is difficult to treat a tuberculous patient who is also suffering from diabetes. When a tuberculous patient is having rest treatment he is given the same amount of food as is needed by a hard manual worker. In the case of a tuberculous diabetic the chief aim of the treatment is to see that the patient assimilates as much food as possible and when we have managed to arrest his loss of reserves and have succeeded in making him take a proper quantity of food we have only to keep this situation under control and then we have an ordinary case of tuberculosis to deal with.

In this connection insulin and a special régime are required.

MEDICINAL TREATMENT

Tuberculosis is essentially a chronic disease, characterised by a variety of symptoms which deeply affect the patient. Fever, cough, sweating, and other symptoms cause the patient to lose appetite and weight and to become less resistant to the disease. The symptom which frightens the patient most is spitting of blood.

In the treatment of the various symptoms medicines should be given only if other means fail. It should be borne in mind that the hygienic and dietetic régime and the other special methods of treatment mentioned relieve the symptoms. If, however, the patient vomits from coughing, a medicine must be given to treat the cough and to stop the vomiting. If a patient loses appetite as a result of fever we must reduce the fever in order to restore the appetite. Ir the symptoms have no effect on the course of the disease and the general condition of the patient, it is not customary to prescribe medicines. It is common knowledge that the great majority of tuberculous patients have been successfully treated without medicines.

Even blood-spitting can be cured without resorting to medicines, but in some cases they are useful for psychological reasons.

If a tuberculous patient is anæmic then certain medicines are indicated.

The general rule must be to give as little medicine as possible to patients suffering from tuberculosis. In every case medicines must be given only under strict medical supervision.

As already pointed out blood-spitting greatly frightens the patient. It is often one of the first signs of tuberculosis, and in itself it is not dangerous, but the patient at once becomes nervous and instead of taking rest he is apt to become agitated. Blood-spitting calls for immediate rest and an ice pack pending the arrival of the doctor who should be sent for without delay.

TREATMENT OF SURGICAL TUBERCULOSIS

Surgical tuberculosis comprises tuberculosis of the bones, joints, glands, kidneys and certain other organs.

The treatment of surgical tuberculosis is based on the same principles as those applied in pulmonary tuberculosis. The same hygienic and dietetic régime is indicated as in the treatment of pulmonary tuberculosis in addition to climatic and light treatment.

Special methods are also employed in the treatment of surgical tuberculosis, the aim being the same as in the treatment of pulmonary tuberculosis, that is to say complete functional rest of the affected organs. Special operations are performed in cases of surgical tuberculosis as in pulmonary tuberculosis, but the types of operations are quite different.

Medical science has made great progress in this field and many forms of surgical tuberculosis can be completely and permanently cured.

Sanatoria and hospitals are the best institutions to deal with surgical tuberculosis. Sanatoria are built at high altitudes in the mountains, and also at the seaside. In the treatment of surgical tuberculosis great use is made of sunshine (Heliotherapy).

INSTITUTIONS FOR PREVENTION AND TREATMENT

For the prevention and treatment of tuberculosis there exist institutions run on sound scientific lines with the latest equipment and trained personnel.

Every country should be divided into tuberculosis regions. Each region should have a complete tuberculosis organisation of its own. In order to nght tuberculosis we need a thorough organisation comprising many different departments. The central point of this organisation must be a dispensary which constitutes the headquarters of the fight against tuberculosis in any particular region. The work carried out here consists in prevention, individual and social diagnosis, and treatment.

The dispensary staff must study the general conditions as they affect the population of the region in which they work. They must have a complete knowledge of the tuberculosis situation in their region (statistics, records, etc.).

It should be made possible for a person suffering from tuberculosis to come of his own accord to the dispensary or he could be seen by the visiting staff who would travel over wide areas in search of tuberculous patients.

People should be examined collectively; school classes, teachers, municipal workers, waiters, factory workers, etc. For the purpose of examination extensive use is made of mass radiography, which renders it possible to examine hundreds of people in a day.

As it is important that there should be a complete record of every case of tuberculosis in the region kept at the dispensary, a law making tuberculosis a notifiable disease everywhere should be passed at once. This is a suggestion which should not shock the public as secrecy will be guaranteed. Patients who come to the dispensary of their own accord or are seen by the staff must receive treatment which should be free. Every person suffering from tuberculosis must be kept under observation or at least should be in touch with the dispensary until he is discharged as cured.

People who have been discharged as cured should still keep in touch with the dispensary for a considerable time. Everywhere dispensaries should be able to function as free out-patients' departments for tuberculosis, as it is sometimes necessary for a patient to be kept under observation for some time before he is sent for a special course of treatment. All dispensaries must if possible function as out-patients' departments besides being centres for the treatment and prevention of tuberculosis.

If a patient is found to be in need of treatment and if his home conditions are not satisfactory he should be sent to the night rest centre.

If a patient has good sleeping accommodation at home, but is not able to work and requires medical attention he should be sent to the rest centre where he may spend the whole day under the same routine as in the sanatorium.

The day and night rest centres are for those whose condition is in the nature of a mild infective form.

If, however, a patient suffers from an active form of tuberculosis he must go to the hospital. Here any appropriate treatment can be given. A patient's stay in hospital is not necessarily a long one and if his condition becomes inactive he returns to the dispensary for discharge or further treatment. The sanatorium is an institution for dealing with curable forms of tuberculosis. Here the necessary operations can be performed and treatment carried out. If a patient cannot be cured he is sent to the isolation home where he is made as comfortable as possible. The isolation home is first and foremost a social institution where hopeless cases are looked after until the end, and where relatives and the public are protected from contact.

A patient who has been in the sanatorium and whose condition has improved is sent to the convalescent home. This is an establishment for patients whose condition is almost inactive but who cannot yet return to work. The next step is to join a colony. Here patients gradually start to work at their own trade or profession under the best possible conditions so that they may in time return to normal life. It has been made clear that pulmonary tuberculosis can manifest itself in many different forms in a patient who may have to pass through the complete organisation before he is finally discharged. But, on the other hand, it is possible for a patient to be sent from the dispensary direct to any of the departments of the organisation according to the form, degree and activity of his condition.

This national or municipal organisation supported by private enterprise, Trade Unions or Friendly Societies, Institutions, etc., is the best way of fighting tuberculosis.

These free institutions should be available to patients and private doctors alike.

A patient can be treated either by a private doctor or in a private or Friendly Society's home, but all records should be sent to the regional dispensary.

These institutions should be the concern of the first department of the dispensary whose main function is individual diagnosis and treatment. (*See* Scheme for Medical Relief Reconstruction.)

The second department of the dispensary deals with social diagnosis and treatment. For this social welfare department a special organisation and trained personnel are required. Their main work is to discover the causes of social diseases and to combat them.

To investigate these social causes a number of mobile units will be necessary. There should also be a mobile unit for the disinfection of homes.

The chief social causes of tuberculosis are malnutrition, bad environment, overcrowding and poverty. The dispensary should have a free canteen to provide food for all who need it. There should also be a department for the free distribution of food. In this department a special section should deal with milk distribution for children, while another department should handle the distribution of

SCHEME FOR MEDICAL RELIEF RECONSTRUCTION

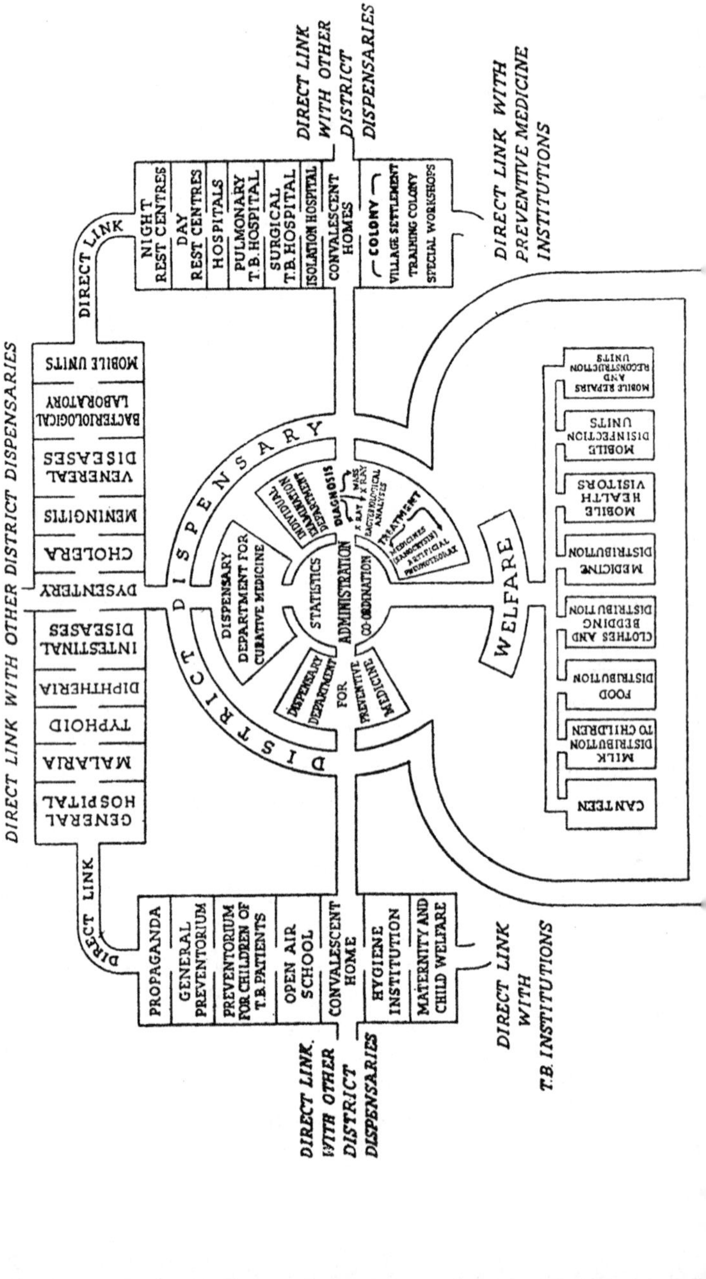

bedding and clothing. A further department should control medical supplies. To make the scheme effective we also need mobile units for health visitors (social diagnosis), for disinfection and house repair work respectively.

The third department of the dispensary deals with preventive medicine. Although the primary object of this department is to fight against tuberculosis it should also be used in the fight against disease in general. This department should control all health propaganda in general and for tuberculosis in particular. Preventoria and convalescent homes should be set up wherever needed. A special preventorium should be built for children of consumptive parents where it would be possible to keep them under observation. Open-air schools should come under the supervision of the dispensary. The different institutions of preventive medicine should also be connected with the general dispensary which in its turn should function in conjunction with the maternity and child welfare department.

The fourth department is the general medical department (clinical and bacteriological laboratory, social diagnosis and treatment). This department should be linked up with the general hospital and other institutions fighting against social diseases such as malaria, typhus, diphtheria, enteric group diseases, dysentery, cholera, venereal diseases, meningitis, etc. To treat the above diseases we need different hospitals. For social work locally such as diagnosis, treatment and disinfection we need a variety of mobile units. They must be equipped with bacteriological and suitable disinfection apparatus and chemical materials.

It should now be clear that the fight against tuberculosis cannot be separated from the greater fight for general health.

The fifth department is concerned with statistics.

It follows from the foregoing scheme that the dispensary is the central point from which the fight for health in a particular region is conducted and in which all records dealing with the health situation of that region should be kept. Every district must have its dispensary. But every district need not have the complete organisation of sanatorium, colony, preventive institution, etc. One large sanatorium, hospital or colony could serve two or three or more districts.

We have described a tuberculosis dispensary organised on broad lines so that it may become not only the centre of the regional fight against tuberculosis but also the central focus in the fight for health in general. The work of the non-tuberculosis department of the dispensary will depend largely on whether or not there are other institutions in the region for the treatment of other diseases and also on the number of people in the region suffering from them.

The central regional tuberculosis dispensary can be used as a model for the central health organisation to be set up in the liberated countries of Europe for the health campaign after the war. As each country or part of a country is freed the social welfare department must be first in the field and for a few months it will probably be working on its own. When the emergency period is over its work will gradually be reduced until it is doing no more than it would do under peace-time conditions.

This dispensary with its complete records of the general state of public health in a certain region would facilitate a thorough survey of the state of health of the whole country by all regional records being sent on to the central national headquarters.

If, at first, we shall need a social welfare department during the early part of the emergency period we shall also need, soon after, special departments for preventive and curative medicine, and tuberculosis institutions. As the general reconstruction gets into its stride and conditions improve the wheel will turn a complete circle and the general dispensary will become the tuberculosis dispensary Many years will pass, however, before we can undo the damage which has been done by the war to the health of the people as manifested by the great increase of tuberculosis morbidity and mortality rates. During the war tuberculosis was the first disease to show a great increase and after the war it will be the last to show a decrease.

Tuberculosis is heading the mortality figures of disease during the present war, and will continue to do so for a long time after. For this reason we must make the fight against tuberculosis the foundation on which to build the whole scheme for general medical reconstruction.

The chief problem in putting this scheme into operation is adequate financial support. Very probably it will be impossible to put this scheme into operation immediately the war is over in all the countries concerned, and in all the districts of each country.

In this connection we should like to mention one particular district in Yugoslavia with a population of about 50,000 inhabitants. In pre-war times under normal conditions the percentage of persons suffering from tuberculosis in that district was 3%. This figure, however, has increased during the war to at least 6%, if not more. This means that in such a district there are about 3,000 people suffering from tuberculosis. Yugoslavia has about 360 administrative districts, some with a higher incidence of tuberculosis than others, but as a conservative estimate it may be stated that there are to-day in Yugoslavia over 1 million people suffering from tuberculosis. The cost of treatment in one single district could be easily estimated. Part of such cost would be borne by the Government as was the practice before the war. As the total sum, however, would certainly exceed the financial resources of the Yugoslav Government the rest would have to be provided by U.N.R.R.A. In so far as it would be impossible to cover the whole country at the same time, the most needy districts should be given priority.

In accordance with the foregoing conclusions we originally put forward the proposal to set up and operate in Great

Britain, first of all one complete dispensary which would serve as a model for many others and which will have to be brought into being for the general medical reconstruction in every country in Europe.

If an organisation of this kind is created in Great Britain any drawbacks or defects will soon be observed and amendments or modifications rendered necessary for the proper functioning of the scheme will be made accordingly. Thus we should arrive at a realistic basis for the creation of a medical relief reconstruction unit.

THE NIGHT REST CENTRE

The night rest centre which was first created in the U.S.S.R. is a department where a patient comes after his day's work for rest and medical attention. On arriving he undresses and has a bath. He then puts on special clothes and goes to the recreation room where he may read or play light games while other patients are seeing the doctor. At the appointed time they all go to dinner. This meal, together with breakfast the following morning, is very important and must contain all the calories needed by a hard manual worker. After dinner the patient returns to the rest room where he may read or play games until bed-time. Between dinner and bed-time those patients who have not seen the doctor are then able to do so. When all the patients are in bed the doctor visits everyone and gives his instructions. In the morning the patients put on their working clothes and return to work.

THE DAY REST CENTRE

The day rest centre is for patients with inactive tuberculosis. The patient spends the whole day there in the fresh air far from the unhealthy atmosphere of industrial towns, and at night he returns home. The journey to and from the day rest centre must be made an easy one. The patient is under the care of a doctor and is treated as if he were in a sanatorium. Patients are admitted to the day rest centre when their sleeping accommodation at home is satisfactory and they do not need full sanatorium treatment. This centre is less costly to run than the sanatorium, it is also cheaper for the patient and the same results are achieved. An added advantage is that the patient does not need to leave his home for long periods. There were two good day centres near Belgrade and Zagreb where splendid results were being achieved before the German invasion of Yugoslavia.

THE HOSPITAL

The hospital admits patients suffering from all forms of tuberculosis, from the most incipient to the most advanced cases. As a general rule patients do not stay in the hospital. Some are sent back to the dispensary to be treated there or sent on to the day or night centres, to the sanatorium or to the isolation home. In the hospital surgical operations are performed and short courses of treatment are given.

THE SANATORIUM

The hygienic and dietetic regime is best carried out in a sanatorium which possesses all the necessary facilities. The sanatorium is built on a carefully selected site. The building is constructed on the most modern lines and equipped with the latest models of surgical apparatus. There is a department for hydrotherapy and light-therapy. Day rest rooms, sun-bathing balconies, parks, etc., are also provided.

Everything is organised so that rest, fresh air, and good food can be provided as required by the régime. Under such conditions medical treatment achieves the best results. The routine is so arranged that the patient lives in well-ordered and peaceful conditions and learns a way of life which will help him when he returns home. This education is of great importance.

THE ISOLATION HOME

Before isolation homes were founded incurable patients were kept in hospitals. Hospitals soon became overcrowded and it was found cheaper to build isolation homes. In order to avoid harmful psychological reactions in the patients the homes were built near the ordinary hospitals.

In Denmark, Norway and Sweden, isolation homes played a very important part. Thanks to them the mortality rate in these countries fell considerably and before the war it was the lowest in Europe. These institutions were maintained and controlled by the State and patients were sent there by law.

THE CONVALESCENT HOME

Convalescent homes are places where patients do not need an elaborate course of treatment but only the hygienie and dietetic régime. They are cheaper to run than sanatoria as they do not require so many doctors to supervise the work nor so much equipment. They should be situated on well-chosen sites.

THE COLONY

In the sanatorium and convalescent home light work is allowed for psychological reasons. Tuberculosis colonies are formed for the purpose of enabling people with inactive tuberculosis to start training for a return to normal life; their earnings from this work help to pay for the maintenance of the colony and money can even be sent to the patient's family.

Colonies are cheaper to run than convalescent homes as they do not need expensive buildings and medical equipment. These institutions have a social as well as a medical importance.

Colonies can be organised to provide work for both industrial and agricultural workers. It is far better to give a patient his normal work than to try to teach him a new profession or trade. The work can be changed if he is unsuitable for it. An agricultural worker is often taken off his work and transferred to something lighter.

Work in colonies is controlled by doctors and all necessary measures are taken to avoid strain and fatigue.

The most important factor from the psychological point of view is that the patient has work to do and no longer regards himself as a sick man. In certain circumstances the patient is allowed to bring his family to the colony and live with them.

The colony is thus a self-contained agricultural or industrial unit where the people live under ideal conditions on hygienic lines. This kind of colony is intended for patients whose condition has become inactive, but in order to be completely cured they must remain under medical supervision for a period of months or years.

PAPWORTH VILLAGE SETTLEMENT

A special kind of colony was founded by Sir P. C. Varrier-Jones in 1916, at Papworth, Cambridgeshire, and is known as Papworth Village Settlement. This village consists of many units. First of all hospitals for treatment of tuberculosis in all its forms. Secondly, chalets where male cases have their quarters while receiving less active treatment and undergoing training. Thirdly, industries together with training of workers and permanent paid employment. Fourthly, the village itself where patients who are able to work and earn money live with their families.

Thus Papworth Village Settlement possesses all the necessary institutions for the treatment of tuberculous patients. The houses are used as night and rest centres by patients and as preventoria by their families. The hospitals fulfil the functions of sanatoria as well as those of hospitals and isolation homes mentioned before. The châlets are in fact convalescent homes. Work in the colony is provided by the village industries. The workers live with their families in small detached houses. The out-patients' department serves as the dispensary. The whole settlement is designed to take care not only of those patients who after treatment will return home, but also of those who will have to stay for many years or even their whole lives.

This is the most interesting settlement in which to study the whole problem of tuberculosis. All forms of tuberculosis are being treated there from the first to the last stages. People with tuberculosis in all forms and stages of development are met there including patients who arrived in apparent good health as members of a patient's family but were later found on examination to be suffering from the disease. In this settlement we have a wonderful opportunity to study tuberculosis in all its stages and aspects. It is interesting to note that very few healthy people become infected there as the precautions taken against infection are based upon sound scientific principles. In the very rare cases when infection occurs this is invariably contracted elsewhere before arrival and the methods of early diagnosis and treatment are so effective that the disease does not make any progress and is soon cured. It is clear therefore that the findings of scientific research are skilfully applied there and an extensive field for research which will enable us to go more deeply into the subject of tuberculosis has been opened up at Papworth.

In his excellent work on "Nutrition and National Health," Sir Robert McCarrison comments on the Papworth Village Settlement as follows: "In this village of 400 persons no child born during the twenty years of its existence has, while a member of the community, contracted tuberculosis of the lungs, bones, joints, cerebral membranes, nor indeed any clinical form of the disease. Yet these children are the offspring of parents who suffer from tuberculosis and are in constant contact with them. How has this remarkable achievement been brought about? Sir Pendrill Varrier-Jones, to whose endeavours it is due, explains it as follows:

"1. Adequate food supply. Ignorance as to dietetic values is dispelled by advice at the clinic, by lectures and by the village nurse. Also by actual demonstration by food supplied from the Central Institution at small cost.

"2. Adequate food supply is possible because there is an adequate and prolonged parental income, maintained by means of assured employment.

"3. Freedom from anxiety as to loss of employment; therefore expenditure can be budgeted in advance.

"4. No risk of unemployment after breakdown; the income being maintained meanwhile: (*a*) in the case of pensioners, by a pension, and (*b*) in the case of non-pensioners, by the Friendly Societies' contributions supplemented by the Welfare Fund.

"5. Proper housing, which allows medical advice to be put into immediate practice; such as through and thorough ventilation in living rooms; isolation of infected persons in bedrooms or verandas. That is to say, avoidance of mass dose.

"6. Public opinion, which makes it possible to live with windows open without being jeered at; to use sputum pots in the house, and pocket flasks out of doors, without being shunned or made conspicuous.

"To sum up: Economic conditions determine the spread or otherwise of disease. To prevent disease it is necessary to create an environment rather than to give a dole where there is no opportunity for money to modify the condition of its recipient. The child's resistance to disease is maintained by: (*a*) adequate nutrition, and (*b*) the absence of mass dose of infection."

It is also worthy of note that Papworth enjoys a better standard of general health than neighbouring villages, especially as regards tuberculosis, in spite of the fact that the inhabibants are near the enemy. This gives point to the saying that if you know your enemy you can conquer him. In taking precautions against tuberculosis people are also safeguarding themselves against other diseases.

Papworth Village Settlement proves that the whole complex organisation we have described can be successfully established in one centre. Papworth is a good example of how the scheme for Medical Relief Reconstruction can be put into operation if we desire to make the fight against tuberculosis the basis on which to build up our general medical Relief Reconstruction Organisation (Scheme). As we have seen, Papworth possesses more than half the institutions described in our general scheme including a large canteen, where good meals can be obtained for the lowest possible charge.

IS PULMONARY TUBERCULOSIS CURABLE?

A person once infected with tubercle bacilli retains for the whole of his life a tuberculous lesion, whether active or inactive. When the tubercle bacilli enter the body they can remain in it for a long time without any noticeable effects. When they enter in sufficient numbers and find a favourable soil for development, lesions will develop. The type of the disease depends upon the number of the germs which enter the body and the nature of the soil in which they take root. Massive infection and favourable soil can cause a severe form of the disease which may prove fatal. Nowadays this is very rare and generally the patient has a good chance of being cured. International statistics prove that the disease can be cured and observations over long periods show an impressive number of cases successfully treated. We can, therefore, definitely state that tuberculosis is a curable disease.

There are very few forms of tuberculosis which from the very first are obviously incurable, a notable exception being "galloping" consumption (Miliary Tuberculosis). The majority of cases taken in hand early are curable.

WHAT WE UNDERSTAND BY CURE AND HOW TO PROVE IT

A complete cure means that the patient does not show any symptoms of tuberculosis and the doctor cannot find any signs of the disease.

A practical cure means that the patient does not show any symptoms of tuberculosis and the doctor cannot find any signs of active lesions.

In each case the bacilli in the sputum are absent.

A complete and practical cure can be proved by clinical X-rays and laboratory examinations. If we compare the findings of early examinations with the successive and the final ones we see that all signs of activity of the disease have disappeared. X-ray photographs clearly show this and serve as documentary evidence. The most convincing proof that tuberculosis is a curable disease is provided by a post-mortem examination of a person who has had tuberculosis and has been cured, and later has died from some other disease or else as a result of an accident. The findings of post-mortems and X-ray photographs enable us to make an exhaustive study of tuberculosis.

While we can demonstrate that the lesions of tuberculosis have disappeared we cannot show that the bacilli are no longer present. This means that recurrence is possible as a result of internal or external infection. The best way of avoiding such recurrence is to maintain strong resistance to the disease by all the means in our power.

IS TUBERCULOSIS AS A SOCIAL DISEASE CURABLE?

We have already stated that tuberculosis is curable as a disease in an individual. The question now arises whether tuberculosis as a social disease is curable. Statistics together with other facts show that it is. It is true that tuberculosis will never completely disappear from the earth, but when it disappears as a social disease it will remain as an individual infection, like pneumonia.

The science of social medicine has shown that as general conditions are improved the morbidity and mortality figures are reduced. Tuberculosis is on the increase wherever the gifts of civilisation are used for the benefit of the few. Hence the greater the exploitation of the people the greater the increase of tuberculosis. On the other hand, if the general standards of living, hygiene, and culture are raised, the death-rate from tuberculosis will be automatically reduced.

Tuberculosis is a useful index of the social and economic conditions of a community. Greed and exploitation must be swept away if we are to eliminate tuberculosis.

Tuberculosis results from the interaction of two factors: infection and soil. Without bacilli there can be no tuberculosis, but without a favourable soil in which to develop the bacilli cannot cause the disease. Social cure will be possible when we are able to destroy the bacilli and maintain resistance against infection. Our ideal is to find a vaccine which will destroy the bacilli in the human organism and at the same time strengthen its resistance to them. The latter is essentially a matter of social reform.

We hope that the time will come when the health of every human being will be guaranteed by the State as a natural right, and safeguarded with enlightened vigilance.

Printed in Dunstable, United Kingdom

91221459R00060